NEW EARTHS

NEW EARTHS
Transforming Other Planets
for Humanity

James Edward Oberg

STACKPOLE BOOKS

NEW EARTHS Transforming Other Planets for Humanity
Copyright © 1981 by James Edward Oberg

Published by
STACKPOLE BOOKS
Cameron and Kelker Streets
P.O. Box 1831
Harrisburg, Pa. 17105

Published simultaneously in Don Mills, Ontario, Canada
by Thomas Nelson & Sons, Ltd.

All rights reserved, including the right to reproduce this book or portions thereof in any form or by any means, electronic or mechanical, including photocopying, recording, or by any information storage and retrieval system, without permission in writing from the publisher. All inquiries should be addressed to Stackpole Books, Cameron and Kelker Streets, P.O. Box 1831, Harrisburg, Pennsylvania 17105.

Opinions and evaluations in this book are those of the author and cannot be construed to represent the views of the author's employers, McDonnell Douglas or the National Aeronautics and Space Administration.

Cover painting and photograph by Dr. William K. Hartmann of the Planetary Science Laboratory, Tucson, Arizona.

Printed in the U.S.A.

Library of Congress Cataloging in Publication Data

Oberg, James E., 1944-
  New earths.

  Bibliography: p.
  Includes index.
  1. Planets—Environmental engineering. 2. Space colonies. I. Title.
TL795.7.024 1981   620.8'5'0999   81-8798
ISBN 0-8117-1007-6   AACR2

I see beyond the range of sight, I hear beyond the range of sound, new earths, and seas, and skies around.

Thoreau

*To my wife
who encouraged the dream—
and to our great grandchildren
who will fulfill it.*

The art on pp. 34, 140, 167, 175, 194, 207, 209, 213, 217, and 254, courtesy of Adolf Schaller. The art on pp. 48, 51, 69, 77, 80, 92, 112, 144, 169, 191, and 211, courtesy Pat Rawlings. The art on pp. 65, 67, 110 (bottom), and 140, courtesy ASTROMEDIA.

Contents

	Foreword by Jack Williamson	9
	Acknowledgments	13
	Introduction	15
1	Terraforming Manifesto	31
2	The Biosphere	41
3	How Earth "Happened"	63
4	Restructuring Earth	85
5	Resources for Terraforming	101

6	The Technology	123
7	Mars—A First Look	147
8	Mars—A Closer Look	174
9	Venus as a True Twin of Earth	196
10	The Smaller Worlds	220
11	Why? Who? By What Right?	242
	Afterword	260
	Notes	262
	Glossary	267
	Bibliography	273
	Index	279

Foreword

Terraforming—the splendid dream that other planets might be transformed to fit them for mankind—once belonged to science fiction. I recall my childhood wonder at the plants that made air for ancient Mars in the novels of Edgar Rice Burroughs. The word itself has always delighted me, for reasons James Oberg reveals.

My own two novels about terraforming began in 1940 with a story idea I sent John Campbell. The idea was for a series of stories about the "planetary engineers" who would one day be reclaiming the moons and asteroids of the solar system. With the creative enthusiasm that made him the greatest editor of his day, Campbell approved the notion and added something of his own: the suggestion that some of the asteroids might be antimatter. A real challenge to the space engineer!

Still new to science, antimatter was then called *contraterrene*. Campbell abbreviated that to *CT*. Spelling it *seetee*, I let its hazards carry me through *Seetee Ship* and *Seetee Shock*, then on to a space-adventure comic strip, *Beyond Mars*, which ran for three years in the New York *Sunday News* (which Oberg recalls reading faithfully as a child).

Now, decades later, I'm wonderstruck again, this time at the way science is overtaking science fiction. When I wrote those stories, the moons of Jupiter and Saturn were no more than enigmatic specks of light in the best telescopes. We know them better now from the Voyager flybys. At the Jet Propulsion Laboratory, watching those stunning pictures coming in, I felt all the thrills of actual interplanetary flight. Watching the *Columbia* climb into space and circle the Earth and glide safely home, I felt it again.

Today, reading the galleys of this remarkable book, I feel that same triumphant excitement awakened once more. New generations of actual planetary engineers have come up with real and dazzling answers to most of the problems that were far-out fiction forty years ago.

Thinking over the breath-taking message of the book, I keep wondering if mankind isn't already making another great evolutionary jump, as epochal as the one that brought our ancestors out of the primal seas a few hundred million years ago, onto dry land.

The first live things on Earth must have been single cells, simple bacteria or their own more primitive progenitors. It seems to have taken them some three billion years to evolve into the many-celled metazoans, able to leave their native seas, but life is changing faster now.

In our own time, new technologies have made the rate of change explosive. I believe we're now in the middle of another long evolutionary leap. Our own new shores, if we get there alive, can be the other worlds of the solar system and even open space itself.

It's a critical point in our history because we can fail. If we do, our kind of life may never have another chance. The survivors of a failure, if any do survive, would be left with resources depleted and their whole environment polluted.

But we need not fail.

James Oberg explores the way to those higher shores and paves it with very solid basic science. His research has been wide; the book is worth reading for its wealth of purely factual data on our neighbor worlds and our own Earth itself, all made admirably clear.

It's worth reading for its daring optimism. Living for a generation in the aftermath of the first nuclear war, beset with unsolved social and environmental problems, too many of us have surrendered to a hopeless terror of technology—a terror so severe that it paralyzes every effort to help ourselves.

New Earths is an antidote we need. It liberates the imagination. It assumes that we can make our own great destiny, that we can save ourselves. It tells us how to complete our great evolutionary leap, off the old Earth into an infinite new horizon.

To some readers the whole notion of terraforming may still seem fantastic, but many things commonplace today would have seemed fantastic fifty years ago. Dry-land life would have been a far-out dream to those early protozoans creeping toward the shore, if they had been able to dream of anything at all.

If we keep science alive and new technologies advancing, they can transform us as well as our planetary environments; they can endow us with powers of thought and action that even science fiction has not yet foreseen. *New Earths* is an inspiring glimpse of what we can do, if only we will.

Jack Williamson

Acknowledgments

The concept of terraforming is such a broad and awesome one that hundreds of people have made contributions both major and minor. As far as this particular book project is concerned, all of the people mentioned in the text deserve thanks. But in addition, there are numerous other people who have made their own unique contributions.

First and foremost are the publishing people who helped convert a manuscript into a book: my editors Neil McAleer, Alison Brown, and Ruth Dennison-TeDesco.

Next are the members of the University of Colorado at Boulder "Mars Project," who took the concept of terraforming and put real numbers onto many of the unknowns, and the NASA team at the Ames Research Center which conducted the 1975 summer study project on rebuilding Mars.

Many members of the scientific establishment have written their own ideas about terraforming or have consulted extensively with me on this project. They include Freeman Dyson, Carl Sagan, Michael Hart, Stephen Dole, Thomas Heppenheimer, Mark Chartrand, Krafft Ehricke, Elbert King, Gerald Mulvey, Everett Gibson, Jr., and others.

Science fiction authors have also contributed ideas, inspiration, and consultation: Arthur Clarke, Isaac Asimov, Jack Williamson, Poul Anderson, Robert Heinlein, Jerry Pournelle, Harry "Hal Clement" Stubbs, Gregory Benford, and Ben Bova. Other authors of stories about terraforming include the late Olaf Stapledon, the late Murray Leinster, Roger Zelazny, and Thomas Scortia.

Special thanks to Fran Waranius and her staff at the Lunar and Planetary Institute Library for aid in research and documentation.

The actual images of terraforming come from artists and photographic services specialists. The artists include Adolf Schaller and Pat Rawlings in particular, along with William Hartmann, Denise Watt-Geiger, David Egge, Don Davis, Ann Norcia, and Mark Paternostro. Photo work by Eric Jacobs and John Waggoner is appreciated. NASA public affairs officers also were crucial in producing the visual material in this book: Michael Gentry, Lisa Vazquez, Don Bane, Pete Waller, historian Lee Saegesser, and space photography expert Richard Underwood.

Many private individuals shared ideas and criticisms with me: Burke Carley (Venusian oceans), Emil Gaverluk (human space destiny), Troy Frantz (demographics), Ed Greisch (Earth rings), J. Richard Greenwell (desert ecology), Frederic Jueneman (geology), William McKinnon (ecology), Liam O'Gallagher (galactic engineering), Rob Quigley (ersatz gravity), James A. Reese (anti-gravity), Carl V. Robertson (soil processes), George Robinson (space colonization), Joseph Suszynski (physics), and Tony Tanis (futuristics). I was also helped in the compilation of a science fiction bibliography of planetary engineering by Richard W. Lair II, R.G. Latimer, Kerry M. Soileau, Joe Suszynski, and Stuart A. Teitler.

None of these efforts, and none of my own would have come to anything but for the patience, encouragement, and understanding of my wife, Cooky. I hope that the result is worthy of the good wishes and optimism of everyone involved.

> *The limits of the possible can only be defined by going beyond them into the impossible.*
>
> "Clarke's Second Law"
> Arthur C. Clarke

Introduction

Terraforming other worlds may take decades to accomplish, and the go-ahead may still be centuries away. So why does such a far-out topic deserve an entire book? What possible value could such a book have?

I believe the time is ripe for such a book, and that the topic of terraforming is worth investigating right now, in the closing years of the twentieth century, and in the still infant years of the Space Age. This topic will be studied more deeply in the future, and this book will be rewritten innumerable times, but a start must be made.

First of all, we now know just enough hard facts about other planets so that our speculations can be grounded in practicality and reality. The atmosphere of Venus has been probed; robots have surveyed the surface of Mars; Mercury, the Jupiter system, and Saturn have been reconnoitered by spacecraft. Speculations have been answered, while new questions have been raised, but enough hard data is right now becoming available to allow us to construct the first tentative scenarios of planetary engineering.

Secondly, a wide circle of people has begun to do this kind of work. Yet, until now, they have been working largely in isolation—alone, "reinventing the wheel," or being stymied by problems other researchers have solved. In the past, people with excellent ideas about terraforming have kept them to themselves, or buried them in a desk drawer because they did not realize that anyone else was interested.

Thirdly, the topic has achieved a level of scientific respectability that shows signs of engendering further studies. For instance, NASA funded a project on transforming Mars. A group of students at the Laboratory on Atmospheric and Space Physics at the University of Colorado at Boulder has been engaged in a broad-based and high quality program to investigate problems of terraforming Mars and Venus. Popular science articles, as opposed to footnotes and half-page fillers, have appeared in magazines such as *Astronomy, Future,* and *OMNI*. The program committee of the Tenth Lunar and Planetary Science Conference in Houston in March 1979 gave its blessing to an unofficial session on terraforming,[1] while cautioning that the topic was sufficiently "speculative" and not sufficiently "connected with current topics in planetary exploration" to warrant a *formal* inclusion in the scientific program.

So the concept of terraforming is a vigorous one, attracting both popular and scientific interest. But why should this be so? Why *should* the possibility of *planetary engineering* (the artificial alteration of the physical and biological conditions of a planet or moon with the ultimate goal of making it habitable for Earth life forms) be studied when it may be centuries before we know enough to even decide if it's a good idea?

Even today, we can imagine several compelling reasons which demand that the possibilities involved with terraforming be studied carefully. First, it's an exciting idea of a possible future that may be complementary to, or an entire replacement of the popular notions of colonies in space. Such thinking, even at an idealized level, has a valuable role in providing options for our future directions. Secondly, the techniques of terraforming will be grand extensions of contemporary techniques in weather and climate control. In an era when human activities have begun to alter Earth's environment on a planetary scale, and when natural forces may well be preparing to make negative changes on our climate, we need to know the mechanics of climate and what affects it. Thirdly, the intellectual excitement of the search for extraterrestrial intelligence ("SETI") requires that we consider all possible manifestations by which such alien civilizations may make themselves known to us. While most scholars suggest that we will pick up their radio transmissions, it is entirely possible that we will detect their existence by noticing the results of their planetary and stellar engineering activities, if we know what such secondary effects might look like.

This book is not an endorsement of planetary engineering, but rather an attempt to examine the factors involved in such activities should that goal ever be seriously established. If any point of view is to be advocated, it is that our options must be kept open, and that the possibility of rebuilding other planets to make them suitable homes for people is a legitimate subject for study and for debate.

Since this book is meant merely as an introduction to this topic and an exposition of the leading problems associated with planetary engineering, there is obviously a need for a more complete technical treatment of the subject. I have already begun work on this project. The book, *Handbook of Terraforming*, will incorporate the world reaction to *this* book, so I solicit critiques and suggestions.

Who Thought of it First?

Perhaps surprisingly, the concept of rebuilding planets is not a particularly revolutionary one; it's been a theme in our literature and mythology for thousands of years. More recently, the idea was given a new name and a new style in science fiction literature, but many of the stories depended heavily on older concepts. In order to better understand cultural development, it will be helpful to trace the development of this idea in the past.

The issues and ideas of planetary engineering will be examined from a number of points of view. First, of course, is an investigation of the term, *Earth-like planets*. To understand that goal, we will have to see how Earth itself was formed and how it evolved, and what random factors combined to create "by accident" the kind of world we propose to create "on purpose." The definition of an Earth-like planet also requires an understanding of the factors which together form a *biosphere*, a planet-wide ecology which supports life as we know it on Earth. Since we are going to have to manufacture an extremely intricate pattern, we need an appreciation of the often subtle ecological chains of cause-and-effect which drive our own home planet's biosphere.

Beyond Earth lie numerous planets, moons, asteroids, comets, and other miscellaneous flotsam and jetsam of the Solar System, bathed in solar energy and containing vast amounts of different kinds of fuels. We will need to inventory these objects and energy sources and classify them by their composition, location, and utility. Next, we will need to examine some of the possible tools and techniques by which the Solar System can be rebuilt; this, in place of the random "hand waving" often resorted to by visionary space prophets. The arsenal of human ingenuity contains far more powerful "tricks of the terraforming trade" than might initially be supposed: rockets, solar sails, thermonuclear bombs and other gadgets, biological and genetic tools, and of course, people themselves.

Mars is the most likely target for colonization, and it has been the topic of numerous recent unpublished studies of great potential significance. So we will consider that planet first, examining its problems and potentials in two full chapters. Venus, the next planet considered for terraforming, has its own special problems, including many not faced by spokesmen who carelessly conjure up visions of air-cleaning algae and similar simple solutions. Besides Mars and Venus are smaller bodies such as the Moon, Mercury, Ganymede, Titan, and others, which are potential sites for constructing new Earths.

Finally, we shall examine the possible motivations, objections, and philosophical issues of rebuilding the Solar System. Those questions also depend heavily on speculations and suppositions which have gone before, so now let us begin by turning to the history of the concept of planetary engineering.

The First Visionaries

The first references to planetary engineering dealt mainly with schemes for altering the spin rate or axis of Earth itself, or with changing the courses of asteroids and comets. These themes were far more widespread than may be realized.

Milton's *Paradise Lost* contains a reference to angels tilting the axis of Earth from straight north-south (which accounted for the mild climate of Eden) to the present 23°, creating alternating seasons. Jules Verne had the same idea only in reverse: his Baltimore Gun Club engineers, having just sent men around the moon, turned to untilting Earth by means of a giant cannon set up on the equator (in order to ameliorate the polar climates so as to allow mining operations). The Russian space genius Konstantin Tsiolkovskiy touched on various topics relevant to terraforming in his works at the end of the nineteenth century, and although his books

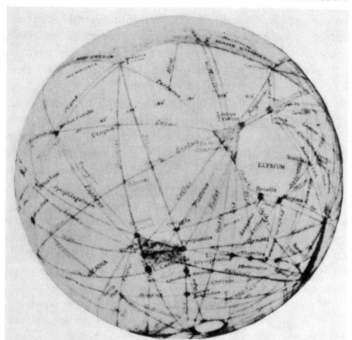

The concept of planetary engineering can be traced back at least as far as Percival Lowell, who imagined a race of Martians building a planet-wide network of irrigation canals in order to counteract an increasingly drying climate.

were not published in time to have any contribution to the work of Western space pioneers, they nevertheless exerted a profound influence on Russian space theorists.

The first alien world to have its climate altered was Mars, at least in the imagination of Percival Lowell. This wealthy Boston astronomer became fascinated with the canals of Mars, and in three books published between 1895 and 1908 he expounded his main thesis: Mars was a dying world, but its inhabitants were engaged in massive planetary engineering to stave off ecological disaster. This fantastic image became the stage for the *Barsoom* novels of Edgar Rice Burroughs, which first appeared in 1912. On Barsoom, giant machinery was used to maintain the atmosphere in a breathable condition. All Burroughs would have had to have imagined was this planetary engineering being conducted by people from Earth, and he would have invented the modern concept of terraforming.

One of the first widespread uses of the ideas of planetary engineering in fiction was in Olaf Stapledon's ambitious "future history" of the human race, Last and First Men. The author used concepts of the alteration of Earth's climate, of a war between Mars and Earth over water, and of an accidental alteration of the Moon's orbit. But perhaps his most evocative contribution was in his description of the moral issues involved with the need for a future race of men to leave a dying Earth and settle a new planet. "Clearly humanity must leave its native planet," Stapledon sets forth as the problem. "The only alternatives were Mars and Venus. The former was . . . without water and without an atmosphere. The latter had a dense moist atmosphere; but one which lacked oxygen. . . . It was necessary either to remake man's nature to suit another planet, or to modify conditions upon another planet to suit man's nature."

Venus was chosen since "Mars could not be made habitable without first being stocked with air and water; and such an undertaking seemed impossible." At first, appropriate vegetation was introduced on Venus (which was covered with a great ocean) to break down the atmosphere and release oxygen; when this proved too slow, electrolysis stations were set up, using some unspecified power source to break down the water into oxygen and hydrogen and the hydrogen was expelled from the planet by "an ingenious method" not described otherwise.

But a problem arose. Deep in the Venerian oceans, a previously undiscovered intelligent civilization "resented the steady depletion of their aqueous world, and were determined to stop it. . . . And as all efforts to parley with the Venerians failed completely, it was impossible to effect a compromise.

"What right had man to interfere in a world already possessed by beings who were obviously intelligent?" Stapledon asked. "On the otherhand, either the migration to Venus must go forward, or humanity must be destroyed. . . ." Additionally, it turned out that the Venerian civilization itself was doomed through climatic changes wrought by natural forces on their own world. "Man would merely hasten their destruction," Stapledon rationalized.

> It was therefore determined to put them out of their misery [caused by man's terraforming changes] as quickly as possible. . . . This vast slaughter influenced the mind of the fifth human species in two opposite directions, now flinging it into despair, now rousing it to grave elation. For on the one hand the horror of the slaughter produced a haunting guiltiness in all men's minds, an unreasoning disgust with humanity for having been driven to murder in order to

save itself. . . . On the other hand a very different mood sometimes sprang from the same sources. . . . As for the murder of Venerian life, it was, indeed, terrible, but right. It had been committed without hate . . . . This mood, of inexorable yet not ruthless will, intensified the spiritual sensibility of the species. . . .

[The project was carried out, but success was less than had been hoped for.] The climate was almost unendurable. The extreme difference of temperature between the protracted day and night produced incredible storms, rain like a thousand contiguous waterfalls, terrifying electrical disturbances, and fogs in which a man could not see his own feet. . . . [A disease appeared which was] finally traced to something in the Venerian water, and was supposed to be due to certain molecular groupings. . . . These troubles were aggravated by the devastating heat. . . .

Over millennia, Stapledon recounted the adaptations to the new homeworld. (Earth, as expected, was destroyed.) Millions of years later, the need arose to move closer to the Sun, so an atmosphere was provided for Mercury. But new factors frustrated that project, and the remnants of the human race migrated to a terraformed Neptune, made habitable by increased solar activity. Almost two billion years in the future, Stapledon conceived of an age when "they gained control of the movement of their planet. . . . They were able, with the unlimited energy at their disposal, to direct it into a wider orbit, so that its average climate became more temperate, and snow occasionally covered the polar regions. But as the ages advanced, and the sun became steadily less ferocious, it became necessary to reverse this process and shift the planet gradually nearer to the sun."

Stapledon's galaxy-spanning prophecies stunned his readers in the 1930's. Even so, they were probably willing to conceive of such occurrences millions or billions of years in the future, but concluded that they had no relevance for the coming decades or even centuries. It remained for subsequent writers to reduce the scope of such visions and bring them within range of our *own* civilization. Planetary engineering was to become a possibility for our immediate descendants.

The word *terraforming* itself, which Robert Heinlein called "a charming neologism, euphonic and self-defining," was coined by Jack Williamson in a series of stories published (under the pseudonym Will Stewart) in the 1940's and republished as the twin novels "Seetee Ship" (1951) and "Seetee Shock" (1949). Williamson described "spatial engineers" who used "paragravity generators" to terraform the Moon, Mars,

Venus, the satellites of Jupiter, and, finally, many asteroids. This was done in the name of human destiny: "The spatial engineers had triumphed over the cold black eternal enmity of space to claim . . . bold new outposts for mankind . . . [and] to cloak in green life all the riven stone of a world born dead."

The first science fiction novel completely based on the concept of terraforming was Robert Heinlein's *Farmer in the Sky* (1950). Three decades after first publication, the tale of the tribulations and triumphs of settlers on a rebuilt Ganymede, one of the giant moons of Jupiter, remains a classic and is in its fifth printing. Newly discovered scientific and climatological facts have demolished the story's scientific basis, but will never alter its drama and its highly readable presentation of the foundations of planetary engineering. (It introduced the word *ecology* more than a decade before the environmentalist movement began to flourish.)

Heinlein's Ganymede was home for several thousand colonists benefiting from a fifty-year-long project which had used nuclear power to melt the ice and ammonia of the original surface and to break down the water into oxygen (which provided the breathing air) and hydrogen (which rose to the top of the atmosphere and hence was "not dangerous"). An unspecified technology maintained a "heat trap" which enhanced an artificial "greenhouse effect," keeping the sunlight in and causing Ganymede to look "like the inside of a sack" from space. But "it's not dark on the ground," Heinlein's narrator realized, since the light got in but just couldn't get out ". . and a good thing, too."

Some obvious and some subtle problems involving terraforming were described in *Farmer in the Sky* (although Heinlein did not seem to realize, for example, that a pure oxygen atmosphere would lead to every piece of organic material on the planet, skin and bones included, bursting into searing flames as soon as the first spark occurred). The crucial nitrogen cycle involved in the biological production of protein was mentioned. The isostatic dangers caused by melting lots of ice and redistributing large amounts of liquid water were involved in the dramatic highlight of the book: an earthquake destroys the "heat trap" which leads to a sudden freezeup which nearly wipes out the colony. The biology is tailor-made: "almost everything grown on Ganymede was a special mutation adapted to Ganymede conditions." Seasonal climate, too, was designed by turning winter "on" and "off" via modulating the "heat trap": "we had to have winter; the freezing and thawing was necessary to develop the ground."

Problems involved with setting up a planetary ecology were not underestimated, as Heinlein's young narrator explained: "The trouble with ecology is that you never know where to start because everything affects everything else. . . . Take the old history book case: the English colonies took England's young bachelors and that meant old maids at home and old maids keep cats and the cats catch field mice and the field mice destroy the bumble bee nests and bumble bees are necessary to clover and cattle eat clover and cattle furnish the roast beef of old England to feed the soldiers to protect the colonies that the bachelors emigrated to, which caused the old maids." This particular chain may be dubious; the concern for such ecological subtleties rests on far firmer foundations.

Social issues also appeared as Heinlein's teenage hero argued with a friend on Earth who was trying to dissuade him from emigrating. The friend said, "My old man says that nobody but an utter idiot would even think of going out to Ganymede. He says the Earth is the only planet in the system fit to live on and that if the government wasn't loaded up with a bunch of starry-eyed dreamers we would quit pouring money down a rat hole trying to turn a bunch of bare rocks in the sky into green pastures. He says the whole enterprise is doomed . . . It's a perilous toehold, artificially maintained and someday the gadgets will bust and the whole colony will be wiped out and then we will quit trying to go against nature."

Of course, Heinlein felt just the opposite, as his character described his own concept of human destiny: "They say man is endlessly adaptable. I say on the contrary that man doesn't adapt himself as much as he adapts his environment." His concept of the essence of humanity allowed him to judge as "human" a vanished alien race because "they weren't animals, pushed around and forced to accept what nature handed them; they took nature and bent it to their will."

The cost of the project brought up an issue which Heinlein could not duck, so he used a very clever way out. Although the public reason for the Ganymede project did involve an attempt to make it financially profitable and to use it for shipping excess population, the 'real reason' in Heinlein's world view was to establish an independent human civilization off Earth, safe from the inevitable nuclear holocaust which population and resource-exhaustion problems would bring to Earth. Enlightened members of the world government were motivated in their support of the expensive Ganymede project by such logic, Heinlein suggested.

Heinlein, more than any other writer, had *humanized* terraforming.

Through his novel he had shown the hopes, fears, doubts, and motivations of realistic people involved in a terraforming project. The change in scale from Stapledon's eons and galaxies to Heinlein's first person narrative coincides with a change in attitude toward planetary engineering. Once it was sheer fantasy for the unimaginable future; now it was *conceivable* for tomorrow or the day after.

Science fiction continued to return to the theme of terraforming. As summarized in *The Visual Encyclopedia of Science Fiction* (1977), Ganymede was still the favorite of many writers. In "The Snows of Ganymede" (1955) Poul Anderson described the terraforming of Ganymede, a popular idea used earlier by Heinlein and more recently in Greg Benford's *Jupiter Project* (1972). Terraforming was the theme of George O. Smith's "Speculation" (1976), which told how organic Earth soil was transported to Mars and cultivated. Earth plants then gradually spread across the planet, the microorganisms in the Earth soil penetrating the Martian surface, paving the way for terrestrial vegetation. The same author's "The Planet Mender" (1952) described cosmic engineering where mountains of ice from Uranus were routed via Mercury and Phobos to become rain on Mars. Heinlein also wrote of cosmic engineering in his book, *Between Planets* (1951). He mentioned a plan to move Pluto and Neptune nearer the Sun, while pushing Mercury farther out. Still another terraforming work, Frank Herbert's *Dune* (1965) depicted a desert planet being reclaimed by its inhabitants.

Scientists Take Notice

References to terraforming were not being made just by science fiction writers. The brilliant galaxy specialist, Dr. Fritz Zwicky of the California Institute of Technology, came out with such forecasts even before Heinlein's *Farmer in the Sky* was published.

Delivering the annual Halley Lecture at Oxford on May 12, 1948, Dr. Zwicky (then fifty years old) remarked that "the application of . . . knowledge to active interference in material celestial affairs and the reconstruction of sections of the universe other than the surface of the Earth has not yet been realized. It remains a distinct possibility for the future . . ." At the close of his paper[2] ("Morphological Astronomy," printed in the August 1948 issue of *Observatory*), Zwicky went into greater detail, suggesting that "In the wake of the realization of large-scale nuclear fusion there will, no doubt, follow plans for making the planetary bodies habitable by changing them intrinsically and by changing their position relative to the sun. These thoughts are today perhaps nearer to scientific analysis and mastery than were Jules Verne's dreams in his time."

Zwicky returned to this theme in later writings. With the dawn of the space age, he considered possible operations on the Moon, while keeping his eyes on more distant horizons. In 1960, he wrote that "modern astronomy intends to embrace two additional activities: the direct experimentation with extraterrestrial bodies and phenomena, and the possible reconstruction of celestial bodies, starting with those of the Solar System. . . . A more complete survey of this future field of endeavor is badly needed." After presenting ideas for the creation of an atmosphere for the Moon, he continued: "For Mars and Venus which are large enough to retain an atmosphere it would thus be a question of altering this atmosphere sufficiently through generation of oxygen and elimination of certain other gases to make them inhabitable. Mars in this respect would seem to be the brightest prospect." Zwicky anticipated that "the chance of exploring entirely new worlds, pioneering in making them habitable, and creating new forms of society, will be the greatest challenge for all great minds."[3] Before he died in 1974, Zwicky had tried to accept that challenge and interest others in it, but without measurable success.

The Venus Algae and Carl Sagan

On the occasion of the launching of the first space probe towards Venus in February 1961 (it failed), a twenty-six-year-old research fellow at the "Space Sciences Laboratory" of the Institute for Basic Research in Science in Berkeley, California, wrote a summary of what was known, suspected, and projected for the planet Venus for the "Current Problems in Research" column of *Science* magazine.[4] The young scientist's name was Carl Sagan.

"It appears very unlikely that there are indigenous surface organisms at the present time," Sagan wrote, since scientists were beginning to suspect that Venus was the hell-hole it actually is. "If, indeed, Venus proves to be without life, there will exist the prospect of microbiological planetary engineering."

Such a speculative topic had rarely been mentioned before, and it was tagged onto the end of the otherwise very sober summary of current knowledge and ignorance about the planet. But there was no attempt to downplay the significance of Sagan's proposal, taken from the pages of science fiction and now proclaimed on the pages of *Science* magazine.

"To prepare Venus for comfortable human habitation, it is necessary to lower the surface temperature and to increase the partial pressure of molecular oxygen. Both ends could be accomplished if a means were found to dissociate carbon dioxide to oxygen and elemental carbon . . . Extensive laboratory experiments should be performed on the ecology of the algae in simulated Cytherian [Venusian] environments . . . The microbial re-engineering of Venus will become possible . . . The greenhouse effect is rendered less efficient and the surface temperature falls . . . Surface photosynthesis becomes possible . . . Rain will reach the surface . . . Venus will have become a much less forbidding environment than it appears at present . . ."

It really doesn't matter that Sagan was quite wrong about what might happen, and far off base on his concept of the conditions on Venus. He had estimated temperatures still too temperate, and surface pressure still too low by a factor of thirty. He believed that "the quantities of water vapor in the two atmospheres are approximately equal," when Venus is actually highly depleted in water in comparison to Earth.

As new data came in, the "algae solution" for rebuilding Venus was appropriately modified. More important than the factual problems, however (and Sagan was as well informed as anyone else at that time), was the public's exposure to the concept of planetary engineering on Venus, and the respectability which the idea garnered from having been mentioned on the pages of *Science*. Terraforming still had a long way to go to acceptability—but it had started along the road.

Taking Planets Apart

Terraforming received additional scientific respectability from Dr. Freeman Dyson of the Institute for Advanced Study in Princeton, who described various types of planetary engineering possibilities in light of their interstellar detectability. That is, perhaps we would detect extraterrestrial civilizations, not from their deliberate radio transmissions, but from noticing the physical results of their planetary and stellar engineering activities.

Speaking in California in April, 1965, Dyson[5] described his approach to what is "possible" for planetary engineering: "My rule is, there is nothing so big nor so crazy that one out of a million technological societies may not feel itself driven to do, provided it is physically possible . . . I assume that all engineering projects are carried out with technology which the human species of the year 1965 A.D. can understand. This assumption is totally unrealistic. I make it because I cannot sensibly discuss any technology which the human species does not yet understand . . . My third rule is to ignore questions of economic cost." On these grounds, what kinds of planetary engineering are possibly waiting out there on the galaxy to be discovered?

"It is possible to take planets apart," Dyson claimed. "One can think of several feasible methods of disassembling a planet." One method, which Dyson describes using Earth as an example (while disclaiming any advocacy of actually doing it to Earth), involves using electromagnetic force to transfer the momentum of orbiting satellites into Earth's rotation, and accelerating the planet to the point where centrifugal force could tear it apart. Calculations show that if all of the solar energy passing through a region three-quarters of a million kilometers across (about equal to the circle bounded by the Moon's orbit) were applied at 100% efficiency to speeding up the spin of Earth, the disintegration point (when the "day" was reduced to less than 100 minutes) would come in about 40,000 years. Dyson had proved his point.

Since at least some technological civilizations would expand out into the galaxy and would engage in massive stellar engineering, Dyson asks why we have not detected signs of such activity. "We must ask: what does a galaxy look like when technology has taken over? . . . A galaxy in the wild state has too little matter in the form of planets, too much in the form of stars . . ." A "tame" galaxy, then, would be different: there would be enhanced infrared radiation, the "waste heat" of futuristic industries; there would be a large ratio of nonluminous to stellar mass; there would be a lot of star-star collisions or artificially induced supernovas, since these are two potential techniques for taking stars apart; there would be "an unusual abundance of short-lived giant stars and deficiency of ordinary dwarf stars," because of the former's utility for supporting habitable solar systems over the "short run" of several hundred million years.

As an example of the contributions such theorizing about terraforming can have for other, seemingly unrelated intellectual pursuits, Dyson then proceeded to tie his *galactic engineering* speculation into the context of the search for extraterrestrial intelligence (or "SETI"). "Why do we not see in our galaxy any evidence of large-scale technology at work? . . . I have the feeling that if an expanding technology had ever really got loose in our galaxy, the effects of it would be glaringly obvious. Starlight instead of wastefully shining all over the galaxy would be carefully dammed and regulated. Stars instead of moving at random would be grouped and organized . . . so in the end I am very skeptical about the existence of any extraterrestrial technology."

"Dyson Sphere" is the term given to one such planetary engineering project described by the Princeton scientist in the early 1960's. Once a large planet had been taken apart (Dyson computes that Jupiter's total vaporization would require "only" the Sun's total energy output for 800 years), the material could be used to construct a sphere around the entire Sun at a distance equal to the radius of Earth's orbit. In some versions, this sphere is solid; in others, it consists of an intricate ballet of separately orbiting small objects. Whatever form it might take, it would nearly fully utilize the entire energy output of the Sun.

NASA Recognizes Terraforming

Terraforming finally received an official blessing from NASA when, in 1975, it was the topic of a special study concentrating on Mars. A team of scientists at the NASA Ames Research Center near San Francisco examined the possibilities of introducing Earth life on Mars, or, if Mars was not habitable in its present situation, of altering the conditions of the planet. Their study was published by NASA in 1976 and was entitled *On the Habitability of Mars: An Approach to Planetary Ecosynthesis*.[6] Several members of the study group, along with unofficial consultants, again met in November 1976 at the 13th Annual Meeting of the Society of Engineering Science at NASA's Langley Research Center near Norfolk, Virginia. They put together a special session at the conference, but were unable to call it "Planetary Engineering" as planned; it came out "Planetary Modeling" instead. The titles of the presented papers, however, gave clear indication of what was being discussed: "The Making of an Atmosphere," "Atmospheric Engineering of Mars," and "Creation of an Artificial Atmosphere on the Moon."

The Public Notices Terraforming

In 1978–1979 there appeared in popular scientific magazines a series of articles on the topic of planetary engineering. First off the presses, in the May 1978 issue of *Astronomy*, was an article by this writer entitled, simply, "Terraforming." Public reaction was good, so the article was rewritten by the staff of *Astronomy* for their children's astronomy offshot magazine, *Odyssey,* appearing in the January 1979 issue under the title "Man on Mars—It Might be You!" The July, 1978 issue of *Analog Science Fiction/Science Fact* carried an article by Ralph Hamil entitled "Terraforming the Earth," which did not mention other planets, but did catalogue a myriad of planet-wide engineering projects which could transform the face of our own planet. Meanwhile, NASA futurist Dr. Jesco von Puttkamer, a regular contributor to *Future* magazine (later, *Future Life*), wrote "In Earth's Image: The Terraforming of Other Planets" for the March 1979 issue. Coming out at practically the same time was an entirely new version of this writer's speculations, published by the new *OMNI* magazine as "Farming the Planets".

The First Terraforming Colloquium

By late 1978 it appeared time to prepare for one more step on the road from science fiction to science speculation. Anticipating the new data which would soon come in from Venus and Jupiter, as well as the continued publication of Viking data analysis, I made inquiries about the suitability of holding a speculative special session on terraforming at the imminent annual Lunar and Planetary Science Conference, to be held in Houston in March 1979. Encouraged by volunteer participants and by the conference staff, a public "call for papers" was announced at the American Astronautical Society conference in Houston on October 30. Over the months that followed, a wide selection of previously uncoordinated research came to my attention. One disappointment was the decision by the conference management to allow the terraforming session to use conference facilities, but not to give the session any official status. According to the letter announcing this decision, terraforming was "highly speculative and not closely linked to exploration of the planets, the theme of this conference." It was not an unreasonable judgment, and the session planning went forward.

The evening of March 16, 1979 marked the world's first conclave of terraforming researchers, who had travelled from California, Colorado, and Washington, D.C., to be with fellow enthusiasts from Texas. Prior to the colloquium, the participants met at a picnic supper on the lawn of the Lunar and Planetary Institute, a handsome building on the shore of Clear Lake which had formerly been a millionaire's summer retreat.

The picnic's idyllic setting was marred by attacks of fire ants, underscoring the small scale problems of setting up a planet-wide ecology hospitable to human life—somewhere else, if not on Earth!

The colloquium itself lasted four hours and drew over one hundred planetary scientists from the Tenth Annual Lunar Science Conference, then in progress. The contents of the papers discussed that evening have been incorporated into this book. Beyond the technical topics was another aspect of the concept which moved one nationally renowned space scientist to proclaim, "I'm most impressed by the enthusiasm of these young people and their willingness to do their homework and face real problems." For much of the scientific community, the colloquium served as a public announcement that yet another "crackpot idea" was about to move closer to scientific respectability at last.

The time was ripe. Decades, even centuries of writers had popularized and prophesied. Independent theorists had gone their own ways in isolation. Now the separate threads could reinforce each other and be woven into a tapestry of stunning perspective which this book attempts to portray.

> Humankind will not remain forever on Earth, but, in search of room and energy, will at first timidly venture beyond the limits of the atmosphere and will then boldly move forth to occupy all the regions around the sun.
>
> Konstantin Tsiolkovskiy

1

Terraforming Manifesto

The first human beings to visit another world beyond this Earth sensed immediately the incompleteness of its Creation. When the Apollo 8 crewmen saw the Moon close up, they seemed to realize that, although Earth had undergone a full evolution from dust to human home-world, the evolution of the Moon was unfinished. It remained sterile and lifeless, an affront to the astronauts' sensibilities.

Consciously, perhaps, they did not anticipate the possibility that future generations of spacefarers would be able to take up where nature had left off, and would be able to remake the desolate Moon into a smaller copy of Earth, complete with its own oceans, forests, prairies, and cloud-studded blue skies. Few people anywhere on Earth, even now, appreciate that remote possibility.

Yet that possibility, and the stark contrast between the oasis Earth and the stillborn world of the Moon, might have been the basis of the insight that moved those astronauts to their unlikely choice of the Biblical

32   New Earths

In traveling to new lands, people have had to bring protective clothing and provisions. But already, humankind has begun to reshape the surface of the Moon. Some day they may be able to hang up their spacesuits in closets and breathe the manmade atmosphere of a new Earth.

reading that they radioed back to Earth. It was Christmas Eve, 1968, but the far-ranging spacemen did not choose to read from any Gospel account of Christ's nativity. They went, instead, to the creation of Earth—and foreshadowed the nativity of a whole new world.

"In the beginning, God created the Heaven and the Earth," intoned William Anders across 400,000 kilometers of space. "And the Earth was without form, and void." Below him, an empty, half-formed world passed by his porthole.

Following the turn of Apollo 8's navigator James Lovell, spaceship commander Frank Borman finished the reading: ". . . And God said, let the waters under the heaven be gathered together unto one place, and let the dry land appear: and it was so. And God called the dry land Earth . . ."

The suggestion that human activity may someday be able to cultivate other worlds, bringing forth vegetation and animal life on a worldwide scale, may in the perspective of the Bible reading appear presumptuous, even blasphemous. Yet it is only a logical evolution of human activity over the millennia, different only in scale from earlier activities.

Rebuilding Environments

Since human beings first huddled around a campfire, or patched together protective garments against sun or cold, they have sought to live in environments for which their naked bodies were not designed. The aqualung and the oxygen mask are modern technological extensions of this basic principle. Philosophers and anthropologists often mark the beginnings of human intelligence at that point in dim prehistory when human beings began to deliberately alter their external conditions or environments, rather than passively accept the physical environment by changing and evolving their physical selves. After that conceptual breakthrough, the march from scraped animal hides and campfires to spacesuits and life support systems was merely one of technology, not concept.

The external environments which have been altered by human activities have grown in scope, from the warmth of a cave to the clearing of a forested valley, then to a regional scale of terraced hillsides, drained swamps, and eroded countrysides. To make life more comfortable, or to open new regions to habitation, people have diverted rivers, tried to make the rain fall, pushed back the sea, destroyed and rebuilt native ecologies, and thrown mountains into marshes. And by accident, unintentionally and in ignorance, human activities have made deserts, wiped out local natural balances, and polluted rivers, lakes, and estuaries.

This philosophy of environmental modification for human gain is bound to continue, on Earth and off. Tempered by lessons learned from unforeseen consequences of environmental abuse on Earth (primarily due to illusions about the unlimited ability of Earth to endure pollution and to provide material resources), humanity can carve out new lands for human settlement, whether on the Matto Grasso or on Mars; the Tibetan Plateau or Venus; the ocean bottom or the Moon; the Gobi or Ganymede.

The other worlds, potential candidates for new homes for humanity, are presently inhospitable to unprotected human bodies. So when human beings first venture there, they will fashion protective garments, and artificially warm their shelters, and carry their own supplies. So it was in the first moments of human history, when the first ventures were made into inhospitable lands; the process continues today with expeditions to the antarctic and the ocean floor. This inhospitality, in space as on Earth,

Since many potential new Earths lack atmospheres and oceans, the necessary volatile material must be imported from elsewhere in the Solar System. Here, "ice-teroids" (a term coined by artist David Egge) from beyond Jupiter smash into the formerly dry surface of one such world—perhaps the Moon or Mars or Mercury.

is not a barrier, only a difficulty. The universe is not actively hostile, merely neutral, and subject (as always) to the consequences of deliberate human manipulation. If the challenges are greater and on a grander scale, then so too will be tomorrow's human skills, powers, and ingenuity.

Rebuilding Planets

Now let us conjure up the visions of planetary engineering. Imagine the terraforming of new worlds in space. When once long ago, fertile valleys needed irrigation, now whole planets lie dry but potentially fertile. Where once caves were warmed, now whole worlds lie ready for thermal manipulation.

Imagine the misnamed "maria" of the Moon filled with water, and the dry rilles doused with torrents from mountain thunderstorms, as the clouds of the new blue lunar skies increase the full Moon's brightness manyfold, casting starker shadows back on Earth. Imagine the pink skies of Mars turned dark blue, and the red rocks tinged with the dark green of algae, as the permafrost buried in a tomb of eons is melted to carve new flood channels and to roam again through the humid skies. Imagine the hell-planet Venus brought to truly deserve its name as the "twin of

Earth," with its suffocating atmosphere locked safely away or dispersed into space, with its months-long "day" repaired, and with a new Earth-like magnetic field gaining strength to ward off solar radiation. Imagine Mercury, shielded and watered, a home for human beings. Imagine the newly probed giant moons of Jupiter and Saturn and Neptune, warmed, sculpted, and watered for settlement.

Imagine the depths of space, 100 million kilometers from any planet, teeming with artificial settlements carrying millions of our not-too-distant descendants. Imagine the foreboding Jovian gas giants, distilled down for their essences, their rocks to form new Earths and their gases to power thirtieth century machines. Lastly, imagine Earth, birth world but no longer prison world of the human race, made deliberately into the paradise people have always sought, but which has always existed only in the eternal imaginations of all-too-mortal minds.

Beyond Earth there lie numerous planets, even more numerous moons, and innumerable asteroids, comets, meteor swarms, and similar debris. Some of these bodies consist of rocky and metallic material, while others are mostly formed of volatile gases and ices. And, of course, there are various gradations of mixtures. All are lit and heated by the nuclear fires of the Sun; some have their own internal heat sources as well.

Through terraforming, mankind will rearrange these physical objects, altering their movements and radiation balances, breaking them up or pushing them together, in order to create Earth-like worlds. Even here the definition of "Earth-like" is still unresolvable; the best we can say is that these rebuilt worlds will be capable of supporting Earth-derived life, including people, without significant restrictions.

The major players in this Solar System reconstruction project will, of course, be the large planets upon which we will try to create habitable environments. Mars and Venus are obvious first choices, with the Moon and Mercury coming along later. Satellites of the Jovian planets have the right size for a sufficiently large gravity field, but their compositions are not always appropriate; such candidates as Io, Europa, Ganymede, Callisto, Titan, Triton, and the Pluto-Charon duet need to be individually considered.

Mars, as the Mariner and Viking probes have shown us, is small and cold with a too-thin atmosphere and an absence of detectable amounts of liquid water. However, the planet is probably the best candidate for the first terraforming target. There is plentiful evidence that great supplies of water are frozen beneath the surface; there are several different schemes for warming the planet to the point that its new climate could be self-sustaining; terrestrial organisms could be seeded there to help convert the atmosphere. We shall examine how all this can be done—along with several technological roadblocks not widely recognized—in two full chapters.

Venus deserves a chapter all its own because of its former reputation as a "twin" of Earth, and its present status as "the best model of classical visions of Hell known in the Solar System." Some of the oldest serious suggestions of terraforming planets dealt with the possibility of rebuilding Venus, but new developments have dealt serious blows to the feasibility of such proposals. While Mars may turn out to be easier to terraform than we had once thought, Venus will probably be that much more difficult.

Perhaps half a dozen smaller, airless worlds are also candidates for terraforming, although wide ranges of solar intensity must be dealt with, and the entire mass of the atmosphere would probably have to be imported. In a chapter devoted to the Moon, Mercury, Io, and similar worlds, we'll look at the individual problems of each, compared to the tools and techniques and cunning which will, by then, be available.

The giant Jovian planets—Jupiter, Saturn, Uranus, and Neptune—consist of tremendous amounts of gases such as hydrogen, helium, methane, and ammonia, together with rocky cores several times the size of Earth. At first glance, it seems their powerful gravity keeps these materials safe beyond human reach—but only at first glance.

This leaves the smaller moons, the asteroids, and the comets. Since they cannot be converted into Earth-like worlds, they can be considered raw materials to be used in the conversion of the more likely targets, the rocky planets. Their primary use will be to provide volatile materials for atmospheres, since such bodies consist (to varying degrees) of water-ice, nitrogen compounds, and similar necessary ingredients. We will need much better knowledge of their compositions in order to decide which ones are the best to use.

Rebuilding Earth

Terraforming has a great deal of promise even if we never consider other planets at all. The climate of Earth itself may need to be preserved against natural disaster by means of terraforming.

For example, it is now generally accepted that our home world is at present deep in an Ice Age, of which we are temporarily experiencing a brief and untypical *interglacial* respite. Climatologists have come to believe that the past few thousand years may have seen the best climate Earth has had (or will have again) in half a million years. Based on predictions of variations in Earth's orbit and rotation angle, the climate here is going to grow progressively worse over the next several thousand years.

This shift in worldwide temperatures is not just a matter of everyone buying thicker overcoats. It is an impending global catastrophe which will see tremendous dislocations of human society. Perhaps—probably—the renewed world freezeup will strain modern technological civilization to the breaking point, leading to economic collapse except on a few tropical islands and subcontinents.

The alternative is for civilization to apply near-future technology to attempt to artificially counteract the climatic shift—in other words, to terraform Earth.

Even then, Earth is vulnerable to other dangers.[1] The latest theory to account for the sudden extinction of the dinosaurs sixty million years ago calls on an extraterrestrial agent: an impacting ten-kilometer-wide asteroid. The cataclysm would have raised enough atmospheric dust so as to effectively shut off most incoming sunlight for several years, resulting in the death of vegetation and of large animals which fed on such plants. After several years, surviving seeds could have sprouted and small animals (who would have endured a diet of carrion and dead plants) would have been the only survivors.

Smaller asteroids have hit Earth and left geological traces (the sauropod object presumably hit an ocean basin). They have "merely" devastated hemispheres, continents, or only regions of a few million square kilometers. Astronomers have calculated that even now there are enough rogue asteroids left over to maintain an Earth-impact rate of one per million years, causing a detonation equal in force to several million megatons of TNT. That's thousands of times greater than the most powerful earthquake possible and thousands of times more devastating than all the twentieth century nuclear arsenals set off together.

And there's more.

Other natural processes, on whose dependability hangs the stability of Earth's biosphere, are now known to be variable. The Sun's own thermal output fluctuates due to internal dynamics still not understood. External factors, such as galactic dust clouds, also seem capable of causing wild excursions from the "norm". Earth's magnetic field varies in direction and in intensity, and it mysteriously reverses its polarity periodically—with disturbing ecological consequences, including the extinction of species. The flux of cosmic radiation can be vastly increased by the appearance of nearby supernovas, bathing Earth's surface in mutagenic and carcinogenic radiation for decades at a time, and hiding underground may not help, since the radiation would completely overturn natural atmospheric cycles involving nitrogen, ozone, and other gases.

All of these changes and disasters have been going on for millions and billions of years, and terrestrial life has apparently repeatedly been subjected to climatic catastrophes. Now Earth is hosting a life form—homo sapiens—whose current civilization is probably not sufficiently elastic to tolerate any repetition of these typical events. If, or rather, when these disasters happen again, civilization could very well be destroyed.

Or maybe not. Homo sapiens is the first terrestrial species to have an alternative; they can learn how to step in and interfere, so as to control or mitigate these natural variations and factors. Human civilization can act as warden to its entire planet, and must do so if long-term survival is to be the goal.

If these ideas seem too far-fetched for serious consideration (in this century at least), they can be shown to be otherwise. As we shall see, rebuilding planets (including our own) is feasible even if we are restricted to technologies and techniques within the range of our late twentieth century imaginations.

As a demonstration, consider some technologies which are really at the outer limits of present possibilities.

Imagine manipulations of subatomic particle physics which allow the conversion of small asteroids and moonlets into miniature suns via the unleashing of runaway nuclear reactions fuelled by the objects' own substance. Arthur C. Clarke and Krafft Ehricke have both independently described such a trick, and they have both realized the uses to which such miniature suns could be put orbiting worlds in the dark, cold, outer Solar System.

Imagine mono-molecular films of fantastic strength, useful for building airtight roofs over celestial objects whose natural gravity is insufficient to long retain a gaseous envelope.

Imagine true artificial gravity—not the pseudogravity of spinning spacecraft where momentum masquerades as mock weight—but authentic inverse-square-law attractive forces which can be turned on and off via the application of energy and ingenuity. Such a device (and its logical consequent, true antigravity) would open the entire Solar System to human occupation, and as an architectural feature of deep-space habitats (each with their own external atmospheres and minature sunlets), such a device could provide homes for the vast majority of our descendants, making living on planets as archaic as living in caves. Earth's surface could be reserved for archeologists, tourists, and a few hermits.

Imagine the opposite of artificial miniature suns—an artificial energy sink—which absorbs radiation indefinitely, providing the means of dissipating radiation belts, supernova blasts, and locally, the heat of the Sun itself so that manned probes could reach and enter the interior of our local hearth-star, or even the interior of Earth itself in search of knowledge and riches.

Imagine distant, unreachable planets (perhaps even other stars) as targets for remote-controlled biological terraforming, as cannisters of spores are launched by a gigantic, interplanetary electromagnetic cannon on voyages lasting centuries or millennia, so that by the time human settlers arrive, the waiting planets will have developed to the point where the colonists can land safely.

Imagine stars themselves as objects to be engineered, first in order to modulate their energy outputs by alternately banking and stoking their nuclear fires, and ultimately as sources of raw material for the construction of Sun-like stars with useful energy outputs and lifetimes (stars could be mined by detonating them or colliding them, and then sweeping up the ashes).

Imagine entire self-contained worlds, stellar systems, or even globular clusters propelled across intergalactic gulfs on voyages of curiosity, conquest, or retreat—with ultimate arrival a million years or more in the future.

Now relax. These stupendous technologies have been deliberately conjured up for several reasons. The first reason is to make tame planetary engineering—the subject of this book—seem less mind-boggling in comparison, and to demonstrate that after a thousand years of terraforming the solar system, our descendants are going to have even bigger projects to tackle. The second reason is to show how the concept of terraforming impacts (often in a subtle way) upon most other topics of modern space technology and theory.

Take "SETI"—the search for extraterrestrial intelligence—as an example. Terraforming is an important topic here. If there really are alien civilizations loose in the galaxy, they are probably engaged in some form of terraforming, perhaps even in a form far beyond those wild imaginings just presented. As space philosopher Dr. Freeman Dyson has expressed it: "We are searching for extraterrestrial technology, not intelligence. We are not interested in what the average extraterrestrial looks like, but only what the most conspicuous might look like."

Such a civilization might make itself deliberately conspicuous by attempts to broadcast recognizable radiation signals across the galaxy (this is the assumption of current SETI theorists); but equally likely, such a civilization might become evident to Earth science via byproduct energy fluxes and modulations consequent to its planetary and stellar engineering activities. Alternately, we might detect—and hopefully recognize—traces of interstellar spaceprobe exhausts, or echoes of thermonuclear blasts of interstellar wars, or something else.

Another example: theorists have wrestled with the question of the origin of life on Earth, and the possibility that such a process has also occurred elsewhere in the Solar System or in the galaxy. Difficulties connected with describing this "accidental" process under postulated terrestrial conditions five billion years ago have led some scientists to wonder whether life might actually have originated elsewhere—under less hostile conditions, and in an originally simpler form—and then have drifted to Earth as spores; this theory has been dubbed "panspermia." A few people have even suggested that the implantation of life on Earth was deliberately accomplished by some extraterrestrial intelligence; this subtheory has been called "directed panspermia." Such an artificial effort to reshape a planet by biological processing is typical terraforming and the implication is startling: the Earth itself may once have been a target for terraforming.

These manifold speculations are beyond the intended scope of this book. Let us leave such stellar engineering topics for later examination and devote our attention to our own Solar System, our future, and new Earths.

> "The thing the ecologically illiterate don't realize about an ecosystem is that it's a system. A system! A system maintains a certain fluid stability that can be destroyed by a misstep in just one niche. A system has order, a flowing from point to point. If something dams that flow, order collapses. The untrained might miss that collapse until it was too late. That's why the highest function of ecology is the understanding of consequences."
>
> Pardot Kynes
> First Planetologist of Arrakis,
> in Frank Herbert's novel
> Dune (1965)

2

The Biosphere

A functioning biosphere is a marvelously intricate machine with an unknown capacity for absorbing punishment. To some extent it is self-healing, but the breaking point may be passed irrevocably long before any external signs of impending, inevitable death can be discerned. Hence, considerably more detailed knowledge of Earth's biosphere is critically important to humanity's future on its home planet, no matter how such knowledge may or may not be applied on other worlds.

Creating such a biosphere artificially on other worlds requires an intimate knowledge of how the biosphere functions on Earth, so that a reasonable subset (perhaps with artificial components replacing some of the organic systems in action on Earth) can be created. This requisite knowledge of Earth's biosphere does not yet appear to exist, so a review of what we do know about the mechanism of this planet can help isolate those questions whose answers are needed both to survive on Earth and to expand onto the planets.

Planetary Connection

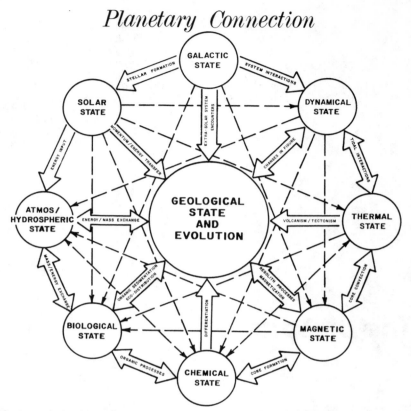

The interrelationships between dozens of factors on a planet are still poorly understood, but must be learned if humankind is to survive on Earth and to set up artificial biospheres on other worlds. *Art courtesy of the Lunar and Planetary Institute.*

An important question for terraforming is to determine how much of Earth's biosphere needs to be duplicated. If it were necessary to mimic the entire Earth, terraforming could be delayed forever. If only a subset would be sufficient, it remains to define how *large* a subset.

Although many aspects of Earth's biosphere seem to rely on massive factors such as the amount of water, or the amount of nitrogen, or the rate of processing carbon dioxide, other factors are more delicately balanced. Good examples would be the ozone layer, the cloud cover over the polar regions, or the ocean chemistry processes which cycle salts into the air and back to the sea in runoff. Such factors may provide fulcrums at which artificial or natural "levers" may be placed to move the entire biosphere off course, so to speak, into a new operating mode.

The Biosphere   43

Many controlling factors in a biosphere depend on materials or energy flows which are remarkably small; modulation of such environmental fulcrums by natural or artificial levers (intentionally or accidentally) can produce effects out of all proportion to the material and energy expended. To coin a metaphor, it is a type of biospherical ju-jitsu, where tiny slaps and shoves can use the momentum of the whole planet to completely change its biological course, without directly confronting the massive energy flows of the planetary climate.

On Earth, such fulcrums may be vulnerable to artificial perturbation to the extent that our planet could be rendered uninhabitable by the manipulation of a proportionately tiny amount of energy or matter. Conversely, on other worlds, a similarly gentle and subtle tickling of such fulcrums may make their environments more benign.

It would be pretentious to claim that a mere chapter can describe in any degree the detailed, complex, and interactive aspects of Earth's biosphere. The most basic factors have been outlined here so they can be considered in the design of other planets. A small appreciation of the complexities and feedback systems of a biosphere will help dissuade would-be terraformers of any easy solutions or "quick fixes" for alien worlds, and will help provide a better understanding of Earth.

What is a Biosphere?

The *biosphere* is defined as that part of Earth's environment in which life actively exists: liquid water can exist there in substantial quantities and can change state between liquid, solid, and gaseous forms; energy is received from outside, ultimately from the Sun; organic machines convert that energy into stored forms, using water in their biochemical factories.

Where sunlight is available, it is used to power the biochemistry of plants; the process, known as photosynthesis, has probably affected Earth's atmosphere more than any other. Photosynthesis can be considered simply as the fixing (or combination) of carbon dioxide and water vapor into a carbohydrate (starch) by chlorophyll-containing organisms, powered by visible light. The oxygen which is given off is from the water molecules.

In general, the reaction is as follows:

$$CO_2 + 2H_2A + \text{light} \rightarrow CH_2O + H_2O + 2A + \text{energy}$$

... where CH_2O is formaldehyde, the simplest organic molecule. H_2A on the left side is normally H_2O, giving off free oxygen on the right side. But there are bacteria that can use compounds in which "A" stands for sulfur, for some organic compound or for nothing at all.

Sunlight Drives Earth's Climate Machine

The "climate machine" of Earth is driven by external energy: sunlight. The Sun bombards Earth with 130 trillion horsepower every second (which is only one half of one billionth of the Sun's total output). Due to the eccentricity of Earth's orbit, this value can vary plus or minus about 3%. The solar output itself may vary by about 1% over years or decades.

Earth's atmosphere introduces much larger variations into local sunlight absorption. The most important factor is the *albedo,* or reflectiveness, of the clouds and of the surface. The scattering of light by particles in the air also lowers the surface temperature. Two more factors which act in the opposite direction and keep in heat that otherwise might be reflected back into space are the humidity of the lower atmosphere and the concentration of gases such as carbon dioxide in the atmosphere.

Local temperatures are also affected by the altitude of the Sun in the sky which depends on season and polar tilt; the length of the day which depends on the same variables; and on the spin rate of the planet. The global range of such temperatures determines the vigor with which the winds blow. The amount of solar energy being absorbed over different points of the globe is thus highly variable. Any local excess or deficit is balanced out, either by an increase in the local temperature; by the vertical or horizontal movement of air or water; by heat exchange between the atmosphere and its lower boundary (water is most active); or by the change of the form of water between the solid, liquid, and gas states. These activities then drive Earth's weather, while defining the limits of its climate.

The Albedo Story

It has long been recognized by climatologists that the most likely way to affect climate on the regional or continental scale would involve some mechanism working on this radiation balance. Since the most important factor determining this radiation balance is albedo, let's look at that first.

The Biosphere   45

Earth's atmosphere reflects about 30% of the solar energy falling on it. Of the remainder, perhaps a fifth is absorbed by the atmosphere, or more accurately, by particles in the atmosphere which consequently warm the surrounding air. Half of all incoming sunlight does make it to the ground where it is absorbed.

Without air, the ground's albedo would be about 10%. However, on an average, half of Earth is covered with clouds, which reflect varied amounts of sunlight depending on type and altitude: high, thin cirrus clouds can reflect up to 20%; altostratus and altocumulus clouds reflect half of the sunlight; cumulus clouds reflect about three-fourths. So the presence of clouds will tend, from the consideration of albedo, to lower the temperature beneath them.

But it is never quite so simple. Russian climatologists note that in high latitudes, because the clouds put a "lid" on escaping ground heat the ground is warmer than would otherwise be expected. At low latitudes, the decrease in surface temperature does occur as expected.

The albedo of the surface itself, as mentioned, averages 10%, but it can vary from very dark (such as water, with an albedo as low as 3%) to very light (permanent snow in the polar regions can have an albedo as high as 90%). Two major variables are immediately apparent: temporary snow and ice cover, and the changing nature of the vegetation cover. Both are subject to wide natural fluctuations and are also vulnerable to artificial perturbations. Both are potentially useful in rebuilding other planets and Earth as well.

Snow varies in albedo from 80–90% when fresh to hardly more than 50% when old. It is, of course, an indication of cold surface conditions. Notice that a snow cover, by reflecting most of the incoming sunlight that would have been absorbed by the ground if *not* for the snow, helps keep the region in the grip of cold. It sets up a self-sustaining *feedback loop:* sunlight drops below normal; land gets cold; snow cover forms; higher albedo keeps land cold even if incoming sunlight resumes normal levels.

This could, of course, lead to *runaway glaciation:* a planet's surface becomes entirely snow-covered, the oceans freeze, and the world remains trapped in ice for the rest of its existence. The albedo is the key operant in this cycle, and in later chapters we will explore terraforming techniques to attack this potential weak point—this planetary fulcrum—on terraforming targets such as Mars.

The recent history of Earth's climate has been related to the nature of the ground cover over climatic regions. As the dominant surface vegetation changes (due to desertification, evolution, or human intervention), the albedo of the surface changes, altering the heat balance and in turn altering the rainfall patterns which in turn seriously alters plant life, thus closing the circle.

The exact mechanism by which human beings have altered ground albedo, and in turn continental climate, will be outlined in the next chapter. Suffice it to mention that ground albedo can vary from as high as 30% over deserts, to 25% for grasslands, to 20% for croplands, to as low as 10% for thick forests, wet dirt, or bare rock. There is wide variation within these figures, and they are generally seasonally dependent, but they have great significance for the whole climate.

Earth's Atmosphere

Leaving the question of surface albedo, consider again the envelope of air which surrounds Earth. This mass of gases interferes with the transport of radiation between the Sun and Earth's surface; it carries heat from one region to another on Earth; it acts as a vehicle for the transport of water vapor and as a stage for the major transformations of water between the gas, liquid, and solid states, which in turn redistributes heat.

Over every square inch of Earth's surface (at sea level) is a column of air weighing about 15 pounds. This weight is the same as 760 millimeters of mercury or about 9 meters (30 feet) of water, and defines a measure of atmospheric pressure called the *bar* (from the Greek for "pressure"). If the whole atmosphere were at sea level compaction, there would be only enough air to reach 9 kilometers (5½ miles) high; however, air density drops only by about a factor of two every 5 kilometers (3 miles) so some air can be found 10, 20, even 80 kilometers high.

The heat which drives the motions of Earth's atmosphere comes from two primary sources. First, a small fraction of sunlight passing through the air warms it directly, but most of the atmosphere's warmth comes from contact with the sun-warmed surface. This results in the lowest portions of the atmosphere being warmer than higher portions, which in turn causes convection since hot air rises. That rising air carries water vapor which, when it condenses high in the atmosphere, releases additional heat which powers the great weather patterns of Earth.

This decrease of temperature with altitude is fairly linear up to a point at the top of the troposphere called the *tropopause*, above which air temperature begins to rise again. The reason for this is that certain constituents of the upper atmosphere, notably ozone, absorb solar ultraviolet radiation and so naturally get warmer. This ultraviolet absorbing layer of ozone rides atop the cooler portion of the atmosphere, blocking the otherwise-deadly ultraviolet rays of the Sun and carrying out yet another important ecological function: the temperature change provides a "cold trap" which prevents the water vapor from diffusing into the upper air where it could escape into space.

This stable ozone layer is not an inevitable result of atmospheric dynamics. On other planets, where the pressure and temperature trends might be different, the ozone might be formed in regions of much greater air turbulence, where violent mixing might frustrate the functioning of the anti-ultraviolet shield or the water-retaining "cold trap." The absolute need for such a layer requires that great attention be paid to atmospheric design for terraformed bodies, particularly smaller ones such as Mars and the Moon.

Wind Patterns

Air is important on a horizontal as well as a vertical scale. It is the major factor in the transport of heat from the equatorial regions (the most sun-lit on Earth) to the polar regions, which may be in perpetual darkness (winter) or in perpetual daylight (summer) where the added sunlight doesn't help much since the snow and ice reflect most of it anyway. The polar regions absorb less than 40% of the solar energy absorbed near the equator. This thermal instability must be relieved by the flow of warm air towards the poles.

Thus, winds are driven. Warm air at the equator rises, replaced by cooler air flowing from the poles. The circulation is not a simple rise and fall pattern, since the equatorial air, flowing poleward at a high altitude, loses its heat and descends to the surface long before it reaches the poles. Several different rise-and-fall stages occur between the equator and the poles, creating wind belts around Earth.

If the globe's surface were not marked by irregular oceans, mountains, and other barriers, Earth's patterns of atmospheric pressure, winds, temperature, and rain would be arranged in east-west belts (as on Jupiter). Instead, we have land topography and water, with their various heating characteristics and their frictional differences.

Still, such an east-west belt pattern is nevertheless recognizable. For

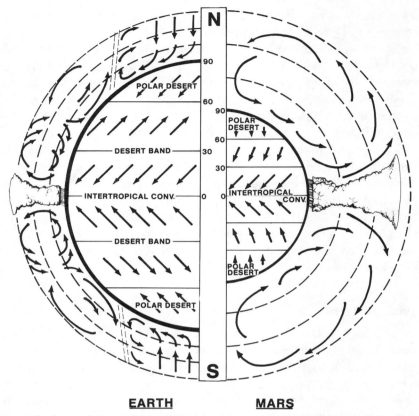

EARTH   **MARS**

Atmospheric circulation depends on planetary size and atmospheric density. The model of Earth's atmosphere is fairly well understood, and scientists are now analyzing data from Venus and Jupiter. Artificial atmospheres on Mars and the Moon must be modeled so that rainfall and ocean current patterns can be predicted.

example, the rising air over the equator carries with it a lot of moisture which condenses at higher altitudes, creating violent rainstorms. This globe-girdling region is called the "inter-tropical convergence," and it fluctuates northwards and southwards depending on the season (its northernmost excursion being called the "monsoon" in India). Meanwhile, this now-dried air at high altitude moves poleward, gradually losing its heat advantage over the underlying air and thus eventually sinking back towards the surface. This dry, descending air creates two symmetric belts of desert regions about 30° north and south of the equator.

In general, the exact location of such desert zones depends on the size of the planet and the density of its atmosphere. Smaller worlds may have only a single circulation cell from equator to pole, and hence their desert zones would be restricted to the polar regions; larger worlds would have many circulation cells and many desert zones.

As air moves away from or towards the equator, it is deflected by the planet's spin, giving rise to the "Coriolis Force." There is, of course, no such thing as a "force" of that name; it's all just angular momentum, making itself apparent in what is more accurately called the "Coriolis effect."

Air at the equator is moving eastwards at about 1,000 miles per hour, the speed of Earth's rotation. A thousand miles north, for example, the eastward speed of the ground is substantially reduced, (since that part of Earth is closer to the axis yet still makes the whole trip in the same "day")—about eight hundred miles per hour. Air from the equator thus has an excess eastward momentum, most of which is lost in surface friction, but a lot of which makes itself visible as an apparent eastward acceleration. It is not: the air is trying to go at the same speed as it had been going at the equator, but the ground itself is lagging!

Whatever the physics, the result is spectacular. The surface winds blowing towards the equator from the north are not purely north winds (that is, originating from the north), but are northeast winds called the "trades." Meanwhile, over the desert zones, the sinking, dry air has no noticeable deflection, but then flows northwards, subsequently being deflected towards the east and creating the belt of winds called the "westerlies;" closer to the poles there is a very unstable pattern reversal giving rise to the "polar easterlies." South of the equator, the pattern is symmetric: southeast trades, a subtropical high pressure zone, and then a belt of westerlies which stretch to the coast of Antarctica.

The strength of these bands of wind depends on the volume of hot air rising in the tropics, and this in turn depends on the temperature contrast between the tropics and the poles. In northern winter, for example, the circulation is vigorous with strong westerlies. The mid-latitude, high altitude jet streams break into eddies which help transport heat and moisture from the tropics to the mid-latitudes.

Hurricanes, too, play a role in this model. These poleward moving cyclones perform a critical function in the transport of heat from the equator; the greater the thermal imbalance, the greater the need for cyclonic storms. In fulfilling this function, such storms also contribute to local precipitation (in some regions, half of the annual rainfall comes during brief and infrequent hurricanes) while influencing evaporation and surface mixing of the oceans.

All of these air motions are, as we said, ultimately driven by heating imbalances, which gives a hint as to where to look for "weak points" in altering climates on a regional, continental, or planetary scale. In the extreme, if solar radiation were not continually being poured onto Earth, the movements of the winds would cease in about thirty days, their momentum dissipated by friction with the ground. This period is also the time in which the air responds to any external thermal imbalance imposed on it.

The Role of the Oceans

Air, of course, is only a small fraction of the volatiles of Earth. All of the weight of air is equivalent to about 9 meters (30 feet) of water, yet there is enough water on Earth to cover the whole surface to a depth of almost 2 kilometers (more than 6,000 feet). So while the air can flow quickly over the whole surface, transporting heat (and heat-carrying water vapor) polewards, the oceans can absorb and release much more massive amounts of heat, over one season or, in the case of the deep oceans, over the centuries. Oceans, therefore, act as a giant "thermal flywheel" damping out any possible wide fluctuations of global temperature in response to transient heat variations.

Surface winds, kicking up waves and whitecaps, influence evaporation (and hence global rainfall) while driving the oceanic currents which drastically alter the climates of coastal regions. Yet all of this is only really happening in the top 100 meters of the ocean. This is the region which sunlight penetrates and warms, below which the water is near freezing all over the world. This is also the layer which moves in the currents, below which the exchange of deep ocean and surface water follows an entirely different pattern on a much longer time scale. This top layer amounts to less than 5% of the ocean's mass; the rest of the water is important, too, but for other reasons.

Earth's inventory of water also shows uneven distribution. Most water, of course, is in the oceans, which amounts to 97% of the planet's moisture. Of the remaining fresh water, three-fourths of it is locked up in ice. The average concentration of water vapor in the atmosphere is enough for a layer 30 millimeters (about an inch) deep, but this concentration varies from about 45 millimeters worth in the equatorial regions, to 20 millimeters in the summer mid-latitudes, to 10 millimeters in the

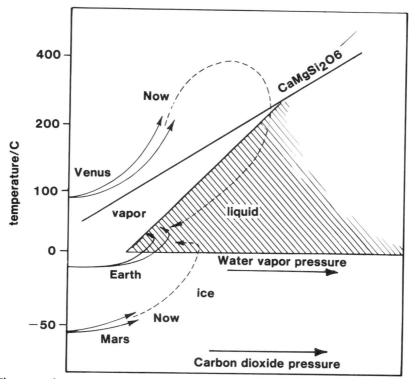

The state of water on a planet is one of the most crucial factors in determining that planet's history. This chart shows how Earth went right and how Venus and Mars went wrong. Artificial activities are needed to push the curves for Venus and Mars into the correct zone in which liquid water can exist.

winter mid-latitudes, to less than 2 millimeters in the winter polar regions. This "trivial" amount of water (about 17 trillion tons of vapor) is a major energizing factor in the planet's weather patterns, and one more illustration of the principle that major climatic effects are often caused by minor constituents, thus providing planet changers with still more fulcrums at which to apply levers for terraforming.

The oceans are the primary source of moisture for global rainfall since the sunlight falling on the water causes it to evaporate. (Only 10% of the absorbed sunlight goes into warming the water.) This water vapor then is transported through the height and breadth of the atmosphere, releasing that energy at various points and providing additional power to the dynamics of atmospheric circulation.

What Causes Rain

Water in the atmosphere is nice for comfortable humidity, but its main utility for biological activity is as a means of transporting and depositing that water over the land—that is, as a means of causing rainfall or snowfall. But just having open water nearby does not lead to rain, as inhabitants of Saudi Arabia or Australia can wistfully attest to.

The formation of rain requires two factors: a cooling of the air causing the humidity to hit 100%, and particles in the air to act as nuclei for the vapor to condense on. There are various ways to achieve these conditions, and human activity has been accidentally and deliberately (some have claimed) affecting these factors.

Precipitation can be classified in terms of the different mechanisms which prompt the condensation; meteorologists list four such mechanisms: *cyclonic, convective, forced convective,* and *orographic.*

Cyclonic rainfall is steady and moderate and is related mainly to the gradual uplift of moist air caused by low pressure zones or wind convergence.

Convective rainfall is shortlived and violent, and generally localized. It is linked with thermal convection causing vertical mixing of air, leading further to thunderstorms.

Forced convective rains are usually tied to particular relief features over which air movement causes steady and prolonged rainfall. Such a mechanism also creates a "rain shadow" on the downwind side of the topographic feature causing the rain.

Orographic rainfall, a more general term which might be considered to include the preceding mechanism, deals with the forced uplift of air over high ground or up a slope. This can include the funneling effect of valleys in upland regions. The rain is steady and prolonged.

Terraformed planets need to be designed with these rain-making mechanisms clearly in mind in order to forecast regions of adequate rainfall and to allocate appropriate plant life to such regions. Precipitation patterns will be intimately tied into biological activity, which will then affect vegetation cover and surface albedo, which in turn effects precipitation.

Superhumidified air still needs material on which to condense, and that material can come from a number of sources. It might be ordinary dust blown from a desert half a world away; it might be biological material dried out and blown into the air; it might be meteoric material from space; it might be salt particles which are left when tiny water droplets from ocean waves evaporate; it might be metallic particles from a storm on the Sun, blown along the solar wind; it might be something deliberately spread in the air by would-be rain-makers. We'll examine sources of such dust in the next chapter.

What is Air Made Of?

So far we have considered the atmosphere purely as a mass to carry heat and dust. Of course, a real atmosphere consists of many different elements and compounds with different properties, ecological roles, and cycles.

Free oxygen and nitrogen are the dominant gases of Earth's present atmosphere, while they are rare or non-existent in the atmospheres of possible target planets.

Carbon dioxide, which dominates the present atmospheres of Mars and Venus, exists in trace amounts on Earth, but is vital to the functioning of the biosphere. Plants use it as a raw material and seemingly cannot get enough of it, since increased concentrations of it merely increase the plant's production of biomass. Animals, however, find carbon dioxide poisonous at concentrations about thirty times above the current level (that is, at about 0.01 bars); the gas pollutes blood and destroys its oxygen-carrying capacity.

Attention should be paid to the potential variability of total atmospheric pressure, which on Earth, of course, is one bar. Different gases which could contribute to an atmosphere have different permissible ranges for human consumption: carbon dioxide, as already mentioned, becomes unsafe above 0.01 bar; oxygen must lie within the range of 0.1 bar to 0.6 bar (the higher density, if not properly mixed with neutral gases, poses a distinct fire hazard); nitrogen, which exists at 0.8 bar now, becomes narcotic at 3 bars. Other neutral gases also have narcotic levels: neon becomes dangerous at approximately 5 bars; and argon, at 8 bars.[1]

That Old Greenhouse Metaphor

The term "greenhouse effect" is widely misused in planetology today, yet the authentic effect is a crucial aspect of planetary climatology.

The metaphorical meaning of the phrase is that sunlight streams through a protective screen (such as glass on a greenhouse, or air over a planet) which is transparent to visible light. This sunlight warms the

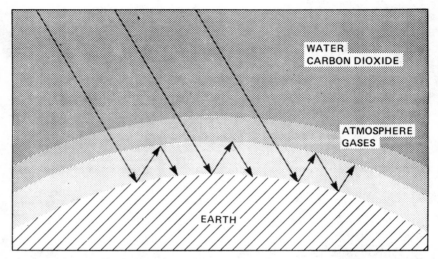

The "greenhouse effect" depends on the presence of certain components of the atmosphere which let visible light pass, but which block out-going infrared radiation. Frequently, the intensity of the "greenhouse effect" depends heavily on atmospheric materials which exist in very small relative abundances, providing yet another ecological fulcrum at which artificial levers can effect disproportionate changes in planetary climate.

surface below, which attempts to reach a heat balance by giving off energy in the infrared. The cooperative screen now balks; whereas it passed the visible light without significant absorption, it is opaque to infrared light—this occurs because of various atomic properties of the materials—and hence absorbs it entirely, heating itself and the surface below it.

But, as it turns out, that's not how a greenhouse stays warm. Panes made of material equally transparent to infrared, such as rock salt, would work just as nicely. The key to greenhouse gardening (which, after all, does work) is that the warmed air inside is trapped by the walls and prevented from mixing with the cold air outside.

And Earth's "greenhouse effect" is also obviously working; our planet's surface temperature is about 40°C warmer than it would be without an atmosphere. Nearly half of the incoming sunlight reaches the surface, where it warms the ground or evaporates water; less than one-tenth of the heat radiated by the surface escapes into space.

The reason for this is that heat is radiated into space at a rate proportional to the fourth power of the absolute temperature of the radiating object, be it the Sun or a desert, or a cloud. (Something twice as hot radiates 2 × 2 × 2 × 2, or sixteen times, as strongly.) Heat from the surface is absorbed by the atmosphere; but the atmosphere itself can radiate heat. However, since the temperature of the upper atmosphere is much colder than the surface (on the average, 33° colder), the heat loss rate is much lower.

Consequently, the atmosphere keeps the heat in. It is primarily the water vapor, and to a smaller degree the carbon dioxide, that contributes the greatest portion of this "greenhouse effect." Terraformers should note, therefore, that this crucial planetary effect is achieved by gases which constitute a tiny fraction of the mass of the whole atmosphere. Manipulation of such mass might be an important fulcrum for planetary engineering.

Oxygen

The role of oxygen in biology is too well known to require any explanation here. However, the gas has another role which forms one more feedback chain and presents yet another constraint on the design of extraterrestrial habitable atmospheres: it supports fire.

The present atmospheric concentration of oxygen is about 20%, or 0.20 bars. Observers might wonder why this concentration, which gradually has been rising over geologic time, does not go any higher. Fire seems to be the answer. If the current amounts of oxygen were to rise in proportion to the inert buffer gases by only 5%, vegetation would become so inflammable that prairies, forests, and even tropical jungles would constantly be catching fire, consuming oxygen, and increasing carbon dioxide and ash particles in the atmosphere.[2] In fact, of course, just such a feedback mechanism is currently operating: forests do catch fire from lightning or from other natural causes thereby keeping an effective upper limit on the amount of free oxygen allowed to accumulate.

This effect is compounded by the fact that land vegetation, so susceptible to catching fire, is a far more efficient producer of oxygen and fixer of energy than ocean vegetation. So oxygen released by oceanic biological activity hits the "fire ceiling" by contributing to land fires, which then cut down on oxygen produced. This fire factor is something would-be terraformers would be well advised to remember. Several of the classical science fiction novels based on terraforming themes have ignored it, and one wonders about the peril to their colonists!

The drifting plant life of the sea, the phyto-plankton, produce about

70% of the world's oxygen (recall that the oceans cover 75% of Earth's surface). Oxygen production on land varies widely, with trees generating much more oxygen than an equal planted area of grass and *most* of the land (deserts, mountains, and frozen tundra) producing little or no oxygen. Yet, however much the land areas produce, they are always in deficit to the oceans: the United States region, for example, produces only about half of the oxygen it uses, relying for the rest primarily on the diatom plankton beds of the northern Pacific.

The life cycle of oxygen atoms is a rapid one. On the average, an atom is recycled from air into a biological material and back into the air once every 2,000 years or so. This tells us that *without* biological activity, all free oxygen could vanish in only a few thousand years.

The establishment of a high-volume oxygen cycle is one of the most critical aspects of planetary engineering. Since we assume that such processing *must* be biological (this may *not* be a valid assumption), this requires the establishment of a complex ecology, perhaps ocean-based. There are a variety of ways to satisfy this requirement without making a complete copy of Earth's biosphere. Some are in this book; some won't be invented for centuries.

Nitrogen

Much of the mass of plants and animals consists of molecules containing nitrogen: protein, enzymes, hormones, vitamins, and nucleic acids to mention the most important. Yet nitrogen requires some very subtle biological processes in order to be included in biochemical reactions.

Nitrogen can link with oxygen to form nitrites (NO_2) which can lead to nitrates (NO_3) which can change to ammonia (NH_3). To recycle this nitrogen, special bacteria are needed—the "denitrifying" brand that chew up nitrates.

Plants, meanwhile, utilize nitrogen via "nitrifying" bacteria that live on roots and which obtain nitrogen from decaying vegetable matter in the soil (or from fertilizers). The material in all living creatures at death is processed by bacteria for conversion into amino acids and other residues, which the "aminifying" bacteria (yet another group) convert into ammonia (named for a temple in Asia Minor where decaying horse manure used to be piled up). Yet another type of bacteria converts ammonia to nitrites, after which different bacteria accomplish the conversion to nitrates so plants can use the nitrogen again.

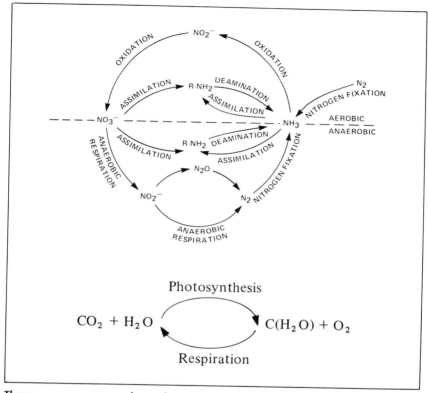

There are numerous complex cycles of recirculation of material in the biosphere, such as this depiction of the biological-nitrogen cycle. If any single step in such cycles is disrupted, the entire biosphere can fall apart. There are dozens if not hundreds of such steps.

Meanwhile, high energy processes in the atmosphere are also feeding the store of nitrates and nitrites by breaking the normally double nitrogen molecule apart, allowing combination with oxygen. These energetic processes, which include solar ultraviolet rays, cosmic rays, and lightning, begin a chain which leads to the washing of these oxides into the sea. For some reason, these products do not accumulate there, probably because some mud-dwelling "denitrifying" bacteria use nitrates to oxidize organic molecules in place of oxygen. This releases the nitrogen to rejoin the atmosphere.

On Earth, nitrogen is in most cases the limiting factor for biological growth (water is important, but it will do no good without nitrogen). Too *much* nitrogen in water causes algae to run wild, and when bacteria process their remains, they consume all the oxygen in the water, killing off the fish. So this balance, at least with the assortment of creatures running loose on Earth, is a precarious one.

To summarize the importance of nitrogen, it must be stressed that sufficient quantities of the gas are required for biological activities, and that a complex cycle of different life forms is needed to keep the cycle open. Without any single link in this chain, nitrogen could be removed from the atmosphere and locked in inaccessible forms. As we shall see, this may already have happened on Mars, where some scientists consider the low atmospheric nitrogen level to be the leading roadblock to terraforming that planet.

The Role of Dust

Dust particles in the planetary atmosphere modulate the heat exchange of the planet and provide condensation nuclei for water precipitation.[3] Although most dust in Earth's atmosphere is natural, manmade dust is already predominant in some regions. Planetary engineers could use dust as a mechanism for regional or global climatic alteration in the near future.

The concentration of dust particles in the air varies widely. This variability affects precipitation, cloud formation and dissipation, and hence albedo in a number of ways. First, additional nuclei for condensation raise the concentration of raindrops in a cloud and reduce the average size of the drops, lightening the impact forces of the rain on the ground and plants below. Second, more dust means an increased concentration of ice crystals in a cloud when the air temperature is below freezing, thereby changing its albedo. Third, particles can give rise to cloud formation above ice-covered regions. Fourth, in principle, particles can deplete the vapor content of the air so the humidity is held down. In practice, the third and fourth items can be disregarded as being practical for Earth.

Beyond affecting the state of water, dust particles directly affect the transport of solar radiation to the surface. They absorb some sunlight, thus warming the surrounding air that would not, without the presence of the particles, have been warmed. (This effect is far more important on Mars.) Particles can cause a change in the radiation field by scattering sunlight, some of it back into space. Although this might suggest that the presence of particles will tend to cool an underlying surface, it's not quite so simple. The actual effect really depends on the optical properties of the particles (including size and color) and of the underlying ground: blacker dust tends to warm, and whiter particles tend to cool the atmosphere.

Most particles seem to come from natural sources, including wind erosion of deserts, soot from forest fires, sea salt, hydrocarbons from plant respiration, and compounds produced from chemical reactions in the atmosphere. Meteoric dust currently contributes little, but its long-term variability is unknown. This also is clearly a mechanism which future planetary engineers might try to use as a climatic fulcrum. Particles are removed by coagulation together, by diffusion to the ground (where dew helps make them stick), and by washout in rain and snow—snowflakes are much more efficient particle scavengers than rain droplets. Particles are continuously being added to and removed from the atmosphere.

The most variable source of particles is volcanic activity and effects of great volcanic eruptions have been recognized for more than a century.[4] When Tambora (on Sumbawa Island, Indonesia) erupted in 1815, it spewed between 30 and 200 cubic kilometers (7 to 50 cubic miles) of dust into the air. The following months saw the "year without a summer" when midsummer snowstorms ravaged New England and Europe, destroying crops.[5] The Krakatoa explosion in 1883 sent 50 cubic kilometers (13 cubic miles) of rock, dust, and ash into the atmosphere, but there were not enough measuring stations to demonstrate the existence of a temperature drop. When Katmai blew up in the Aleutians in 1912, however, solar radiation was measured at 20% below normal in Algeria. Such clouds last one to three years typically.

Following one eruption in 1915, there was a remarkably long spell (several decades) without additional major eruptions, and the particle concentration in this planet's air fell markedly. The warming of Earth's temperature which was recorded over that period may actually have been due entirely to that lack of new sources of particles (however, since 1945 a cooling trend set in which seems connected with some other effect, possibly manmade). Finally, eruptions resumed with Agung (Bali, Indonesia) in 1963, Taal (Phillipines) in 1965, Awu (Celebes, Indonesia) in 1966, Fernandina (the Galapagos Islands) in 1968, and Mount St. Helens in 1980. These major eruptions have again injected large amounts of dust into the atmosphere of the northern hemisphere; since there is no significant air mixing between the northern and southern hemispheres, the air in the southern hemisphere is significantly freer of such dust, and no apparent increase in dust has been detected.

For terraformers, the significance of all this dusty talk is that volcanoes are not the only potential source of atmospheric dust. All we need to do is to generate some correlation between actual mass of dust and actual atmospheric turbidity, and then we can compute the size of a pulverized asteroid needed to alter the mean planetary surface temperature by a given amount. Then, of course, all we have to do is go find such an asteroid and aim it for our target planet, pulverizing it completely on the way.

Life Cycles: Tides, Day, and Night

A major factor in the life processes of coastal species is the daily rise and fall of the ocean surface: the tides. In some continental shelf areas, local topography can force these height differences to be as great as 16 meters, but in fact the actual tidal motion of the seas is quite small. In central ocean areas, such as Tahiti, the tidal range is as little as half a meter (the actual flexure of the "solid" Earth is about a meter in extent); as mentioned, the presence of land masses complicates and magnifies this effect. Other planets without large moons will by necessity have smaller tides, but the Sun's gravitational pull will still provide cyclic water height variations, probably sufficient to support biological processes which depend on them.

Another little-appreciated pattern is the light/dark cycle of 24 hours, induced by Earth's 23 hour 56 minute spin and by its orbital motion around the Sun. Besides its direct effect on photosynthesis, it creates temperature changes of varying severity: on bare ground, the day-night range can amount to several tens of degrees Centigrade; below a depth of half a meter this change is hardly noticeable; at sea, the temperature change in the top 100 meters of water is usually less than 1°C.

It is still unknown just how flexible Earth biota can be for longer day-night cycles. Russian scientists have been experimenting with crops grown on a lunar 30-day cycle. The "day" on Venus is 118 of our Earth days long, which may require plants to follow an arctic pattern of months of sunlight followed by months of darkness. Such limitations have not been investigated thoroughly.

The Role of the Magnetic Field

The role of Earth's magnetic field in the life processes of terrestrial organisms has not been widely appreciated. This magnetic field does more than align compasses and deflect the charged particles from the Sun into Earth's polar regions causing the Northern Lights. It seems to have some subtle but vital connection with the very essences of living creatures.

The effect of the absence of Earth's magnetic field can be studied in two ways: first, geologists can study the records of life during the periodic reversals of Earth's magnetic field, when the field appears to temporarily fall to zero before building back up in the other polarity; second, experimental specimens can be placed in antimagnetic chambers which shield out Earth's field. Both cases provide disturbing data.

Geologists now realize that magnetic reversals of the past few million years have each been accompanied by the total extinctions of several species of marine microflora. Dr. Ian K. Crain of the Australian National University believes that the cause-and-effect linkage is direct: "Mass extinctions are caused directly by the deleterious effects on organisms of the reduced magnetic field during a reversal."[6]

According to published accounts of antimagnetic chamber experiments, the effects are visible at all levels of living things from plankton to lower mammals. Bacteria show a fifteen-fold reduction in reproductive rate. Flatworms, protozoa, and mollusks exhibit different movement, and the motor activity of birds is dramatically altered. Mice suffer drastic changes in their enzyme activity, in breeding, and in reduced life spans.

The reasons are still obscure. Perhaps biological molecules are lined up by magnetic fields, facilitating their interaction. Perhaps some electromagnetic process of the cell is disturbed, such as the transfer of charged ions across the cell membrane. Something seems to interfere with the biochemistry.

In the absence of a planet-wide magnetic field, perhaps an artificial localized field will suffice for biological activity on terraformed worlds. But without a better understanding of the cause of this magnetic connection, no reliable estimates can be given. Whatever it is, it is very subtle since the maximum field on Earth is still several hundred times weaker than the field between the ends of a toy horseshoe magnet.

Our reliance on the existence of a magnetic field would not be nearly as frightening if we had a good understanding of what factors inside the planet give rise to such a phenomenon. There are some good guesses and some promising leads, but geophysicists are still trying to understand the cause of long-term strength changes, of alterations in the direction and pattern of planetary magnetic field, and of sudden magnetic reversals which seem to occur every million years or so.

Earth's magnetic field must be continuously regenerated, since the interior is much too hot for any magnetic material to retain some postulated ancient permanent magnetism that was left over from Earth's creation. Instead, the planet can be viewed as an electromagnet, perhaps driven by gravitational energy liberated by the migration of heavier metals to the center of the core and of lighter metals into the outer regions of the core. This energy may drive helix-shaped eddies in the molten nickel-iron portions of the interior, producing the field and the connected slow changes of the field.

The field is tilted at an angle of only 11° to the axis of Earth's rotation, which creates an obvious and striking correlation between the rotation and the field. Several blind alleys in electromagnetic theory and in geophysics were followed earlier in this century, attempting to conjure up new laws of physics which accounted for this connection. Nothing came of it, and geophysicists turned to other avenues of speculation.

At present, the gravitational energy plus helical ripples theory is the leading candidate. However, as expressed by geophysicists Charles Carrigan and David Gubbins: "No one has developed an explanation of why the sign reversals take place. The apparently random reversals of the Earth's dipolar field have remained inscrutable."[7]

It also was pointed out in *Science* magazine that "Only the Earth of the inner planets has a massive satellite, which may account . . . . possibly, for its strong magnetic field. . . . At least one theory of the Earth's magnetic field, which is much stronger than those of Mars and Venus and which screens the Earth from the solar wind particles ejected by the Sun, depends on the dynamic influence of the Moon's presence."[8] Since the only other planets known to have a powerful magnetic field are Jupiter and Saturn, which also have very large satellites around them, this observation may be a hint of something significant—or it could be just a coincidence. But if there is something to it, it does not bode well for the need of artificial instigation of magnetic fields in terraformed planets—unless large satellites can also be emplaced.

To summarize, we have highlighted the leading factors in a biosphere and have touched on some of their most obvious interactions. It has been a blueprint, so to speak, for a habitable planet. The next question is how this intricate ecological mechanism came about on Earth, so that would-be planet builders can plan how to duplicate the process—or at least aim for similar results—on other worlds.

> *Who knows for certain? Who shall here declare it? Whence was it born, whence came creation? The gods are later than this world's formation; who then can know the origins of the world?*
>
> Rig Veda, X 129

3

How Earth "Happened"

If Earth came about through a series of circumstances different from other planets, then an aspiring terraformer needs to know the detailed history of Earth's formation and development, and the ways in which it differs from its neighbors. Furthermore, such a study may reveal directions along which the evolutionary pattern of other worlds needs to be deliberately directed in order to produce more Earth-like results.

According to Nigel Calder's book *The Weather Machine* (1974), there were five possible ways Earth might have turned out: one, like Venus, a "runaway greenhouse effect" with temperatures of 300°C; another, an entirely ice-free world; one, slightly icy (as now); one, perennially ice-covered (like the deepest epochs of the Ice Ages); the last, like Mars with a "runaway glaciation" deep freeze and temperatures below $-40°C$.

How and why were we lucky enough to stumble onto the one path leading to a hospitable Earth? We still don't know if it was blind luck, ecological feedback (the *"Gaia Hypothesis"*), or something else.

Many of the things which happened to Earth over its history, since they led to our present conditions, are things we hope have also happened to terraforming target planets. If these things did not occur on other planets, it is going to be up to us to *make* them happen there.

How did they happen on Earth?

In the Beginning

From statistical studies of nearby stars, astronomers have deduced that the process of stellar formation involves a collapsing gas cloud which forms into, not one, but a family of bodies. In some cases it is a double star of nearly equal proportions; in others, a star with a dark companion too small to ignite nuclear fusion; in still others (about 20% of the total), a star with planets or unconsolidated debris. Each of these multiple systems is created in a large nebula which is supporting the simultaneous birth of several hundred such systems.

Once formed, these star systems drift apart. Occasionally, however, two nearby systems remain attached through gravitational attraction, forming widely spaced binary or triple systems. The nursery clusters otherwise disintegrate as the galaxy rotates several times, until each small system is separated from its stellar siblings.

Just what sets off the initial condensation in these dust clouds is still a mystery, although a shock wave from a nearby supernova detonation is a likely cause. Even today, astronomers are able to find dust clouds which apparently are the locale of stellar births.

The dust from which our Solar System formed was not primordial material from intergalactic space. Instead, it had been strongly mixed with heavier elements which had been produced in the interiors of older stars, and then had been spewed forth into the galaxy by stellar explosions. So our own Solar System is at least a "second generation" star system, built from the ashes of other stars billions of years older.

In orbits around the Sun, whorls of condensing material formed into small subplanets which then combined gravitationally, releasing energy when they collided, and also heating up from gravitational contraction. During this time, the energy of the Sun was preferentially evaporating more volatile materials from the inner Solar System, since temperatures did not drop to their condensation points until the Sun's heat had been diminished by greater distances. Also, the inner planets underwent pro-

How Earth "Happened"   65

cesses of differentiation, during which molten regions would separate out their materials, light slag rising to the top, and heavy metals sinking in toward the center.

What is Earth Made Of?

Earth is a rocky planet with a dab of volatile material (water, carbon dioxide, nitrogen, etc.) added. By weight, these volatiles only contribute .01% to Earth's total; if gathered together into a frozen mass, they would form a super-comet about 800 miles in diameter. Yet their presence has transformed Earth and allowed life to flourish.

Where did this type of material come from? Traditional geology holds that these materials were released from the rocks of Earth through an *outgassing* process from volcanic vents over the eons. If so, geologic studies of other planets should enable terraformers to determine the inventory of available volatiles there. Opposed to the outgassing school of

Volcanoes erupting on newborn planets probably contributed most of their atmospheres in a process called "outgassing." Here, Mark Paternostro conceptualizes the eruption of the Tharsis volcanoes on Mars. Similar processes took place on Earth, the Moon, Venus, and probably Mercury. Different types of volcanoes are active on Io.

thought is a minority viewpoint which suggests that Earth's volatiles (or a large fraction thereof) are of external origin, results of the impacts of giant comets early in the formation of the Solar System. This hypothesis, too, is of interest to terraformers, since the artificial impacts of volatile-rich objects onto dry planets is one potential technique of planetary engineering.

As far as the outgassing theory is concerned, the composition of the gases given off by volcanoes is a subject of some dispute. Even when it has been measured, the significance of the measurements has been questioned. One of the key issues involves the proportion of "juvenile," or never-before-exposed water.

Water vapor has been measured as a major component of the fumes given off by many active volcanoes, but the degree of ground water contamination has always been in doubt. Water vapor in volcanoes, along the edge of downthrusting geologic plates (such as the Andes chain of volcanoes), can be explained as recycled ocean water trapped in the rock and thrust deep into the planet, where it joins with upwelling magma.

Volcanic activity in the stable interiors of Earth's plates seems to be significantly different. Along the African rift, for example, the lavas are highly deficient in water but rich in carbon dioxide. Such carbonate lavas (typical of such volcanoes as Oldonyo Lengai in northern Tanzania, Nyiragongo in Zaire, and Eburru in Kenya) are essentially anhydrous or water-free. Carbon dioxide, meanwhile, leaks out of the ground all along the great rift, occasionally with such pressure and purity that it can be commercially tapped.

So nothing observed today can really determine the composition or the rates of outgassing four billion years ago. Water vapor is just one of the problems connected with the origin of Earth's atmosphere. While it has been reasonable to assume that such material came from outgassing from the interior of the planet, other sources for atmospheric volatiles *are* possible.

Fred Hoyle, renegade cosmologist and erratic science fiction writer, is one of a minority of scientists who believe that most of Earth's volatiles were originally imported—that is, they originated elsewhere in the Solar System. "From where, then, did Earth acquire its volatile materials?" he asks. "It seems more profitable to seek an origin for the volatile materials in the region of the outer planets, rather than directly from the Sun."[2]

An explanation of how such foreign material could find its way to the vicinity of Earth is not difficult. Two scientists from the Lunar and

An unknown but possibly highly significant portion of the volatile material making up Earth's atmosphere came from impacts with large comets, as pictured by Adolf Schaller. Such a process is no longer occurring naturally, but it can be started up again by human activity where it's needed to provide new atmospheres.

Planetary Laboratory of the University of Arizona have devised a scenario where such a development is possible, even likely. According to Godfrey T. Sill and Laurel L. Wilkining,[3] the gravitational forces of the outer planets, particularly Jupiter, would scatter the cloud of *planetesimals* from which Uranus and Neptune were forming, sending some deep into space to become comets, and sending others towards the Sun and the inner planets. By the time the outer planets had condensed to the size needed to deflect the leftover remnants, even the remnants would be several hundred kilometers in diameter—more than enough to contribute a significant fraction (even a majority) of the volatiles which went to make up Earth's atmosphere and hydrosphere.

Two aspects of this still unsettled question are of particular interest to would-be terraformers. First, it always helps to have an accurate inventory of the volatiles already emplaced on a target planet in various forms, and to do this, speculations must be made about the planet's history and, more specifically, the history of its atmosphere. Second, the concept of a planet's atmosphere having an external point of origin may or may not be valid for the history of the Solar System up until now. However, it is well within the sphere of possible human activities over the next few centuries to deliberately steer volatile-rich asteroids from the outer Solar System toward collisions with dry inner planets, thus fulfilling speculations about such events in the past.

The Importance of Water

The presence of large amounts of water is much more important for a planet's development than simply as a veneer on the surface to support biology. If a planet possesses water in quantity, the substance permeates the planetary crust and grossly alters the physical processes of that planet, both on the surface and deep within the mantle, hundreds of miles below the surface.

Eons later, even if all of the water is gone, these physical traces remain indelible. The very mineralogical structure of the rocks will testify to the former presence of water.

Alternately, a planet which never possessed significant amounts of water will have a geology which is markedly different from Earth. Even adding water artificially, as we will soon propose to do in terraforming, will not immediately create Earth-like surface conditions.

On the largest scale, water supports tectonic activity, or *continental*

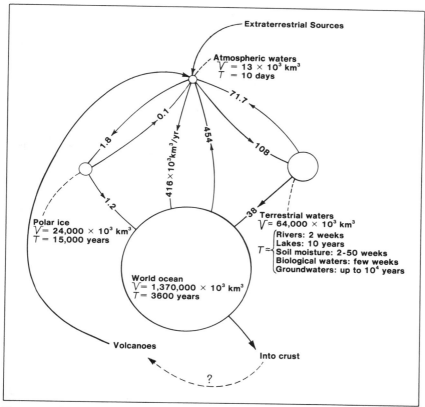

The water of Earth is distributed through several different reservoirs, comes from several sources, and disappears into several "sinks." The fraction of the water which exists in the form of vapor in the atmosphere has an influence on climate out of all proportion to its relative mass. It thus forms a possible pressure point for the application of artificial levers for planetary engineering.

drift. That is, the presence of water intermixed chemically with a planet's surface rocks to a depth of several hundred kilometers lowers the melting point of those rocks and allows them to flow. As the drifting plates crash into each other and force material back into the depths of the planet, more water from the surface is mixed into the deep rocks. Such cycles probably take a hundred million years or more to complete.

Without water in the surface rocks of a planet, the continents would stop drifting and remain fixed over the deep interior layers of the planet. A significant piece of evidence relative to this different tectonic state on other planets is the existence or absence of giant *shield volcanoes*—

eruptive structures built up in one place over hundreds of millions of years—which can reach immense proportions.

On Earth, for example, the presence of continental drift implies that the upwelling of lava from "hot spots" deep in the interior of the planet will poke through the plate at different points as the plate slowly drifts overhead. The result, best typified by the Hawaiian Island chain, is a *string* of volcanoes decreasing in age from one end of the chain to the other. On Mars, however, the Tharsis Volcano is an indication that the upwelling lava from a deep "hot spot" *kept coming out the same hole* for up to a billion years, and clearly suggests that there is *no* continental drift on Mars. Perhaps the interior never got hot enough, or perhaps the melting point of the rocks was not reached because the Martian crust is not saturated with water. Similar indications are given by the Beta Volcano on Venus, which also appears to be a giant caldera. Even though there *are* other indications of some tectonic activity on Venus, Beta tells us that water in the crust is probably not plentiful.

Water also dissolves minerals in the crust and redistributes them. It is not widely known, for example, that it was only through the action of water that mineral concentrations (our precious ore deposits) ever formed from the generally dilute mix of materials which made up the original crust of the planet.

On a planet's surface, water is the dominant mechanism of material transport. It erodes soil from some regions and redeposits it elsewhere, either in river valleys or in sediments on sea bottoms, from which it is lifted by the tectonic activity its very presence makes possible into new continental surfaces. It also dissolves salts and leaches them out of the ground in some areas, making the soil desirable for biological activity, simultaneously concentrating the salts into thick, localized deposits elsewhere.

Such processes take millions of years. The sudden appearance of large amounts of water on a planet hitherto lacking the substance is not likely to initiate continental drift (at least not for hundreds of millions of years) nor is it likely overnight to create mineral concentrations, salt-free soils, sediments, and other water-caused formations with which we are familiar (and upon which our life cycles depend) here on Earth. Water has left its mark on Earth, making it into a habitable planet; without a billion-year history of water's activity, there are serious questions about whether or not any other planet can even approximate terrestrial surface conditions.

History of the Atmosphere

The early atmosphere of Earth (whatever its source) was totally inhospitable to present forms of life: there was no free oxygen, while carbon dioxide and water vapor probably made up the bulk of its composition; and there were traces of ammonia, methane, and nitrogen. Yet somehow over four billion years the atmosphere was transformed into the one we have today.

It has generally been assumed that free oxygen began to accumulate about 1800 million years ago, created by living creatures who still hid from the Sun's ultraviolet radiation by living in the seas.[4] When oxygen reached about 1% of its present level, it would have begun to give rise to an ozone layer sufficient to screen out much of the worst ultraviolet radiation from the Sun. This allowed life to move onto the dry land somewhere between 1000 million years ago and 600 million years ago.

The assumption that all of Earth's oxygen is formed from the biological breakdown of carbon dioxide seems a reasonable one. Rough estimates of the amounts of carbon and oxygen in the biosphere seem close to the expected ratio of 12:32 (carbon has an atomic weight of 12; oxygen is 16; and since CO_2 has two oxygen atoms the oxygen weight

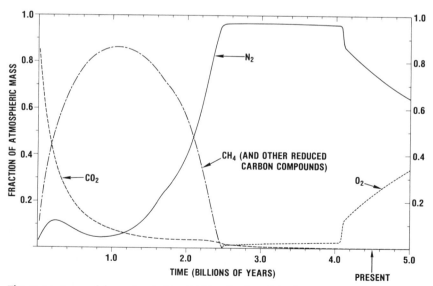

The computer model constructed by NASA scientist Michael Hart provided a plausible scenario for the evolution of Earth's atmosphere. Surprisingly, Hart's study suggested that Earth has more than once teetered at the knife edge of climatic catastrophe.

is 32). Some estimates put the total amount of oxygen released by photosynthesis as high as 32 million billion tons, of which a quarter went into the rocks, a negligible amount escaped into space, a large amount oxidized the components of the prebiotic atmosphere, and a bare 4% remains today in the atmosphere.

Studies to determine the interaction of all these factors in the early history of Earth's atmosphere have been stymied by the complexity of observed processes alone, without considering any unobserved processes—that is, at least, until the computer. A numerical simulation of Earth's development was carried out from 1976 to 1978 by Dr. Michael Hart of NASA's Goddard Space Center, in order to find plausible "initial conditions" and processes which would lead to the present familiar state of Earth's atmosphere.[5]

It may be instructive to examine the different factors which Hart entered into his computer model of the history of the atmosphere of Earth. Those factors apply to planets in general, whether the changes are natural or artificial.

Actions of volatiles in the atmosphere were charted by Hart using a simulation of outgassing rates for argon, carbon, oxygen, hydrogen, and nitrogen. The behavior of the water vapor was simulated using measurements of the planet-wide humidity and the boiling and freezing points of water, based on atmospheric pressure. Additionally, the breakdown of water into hydrogen and oxygen, and the subsequent escape of the hydrogen into space was also simulated.

One *sink* (a place where a material is lost) for oxygen was in the oxidation of surface materials. Hart noted that most iron in the old igneous rocks on Earth is in the ferrous, or non-oxidized, state, while the iron in sedimentary rocks (which formed later) is more likely to be in the ferric, or oxidized, state.

Carbon dioxide had a sink as well, in the *Urey Reaction*, which describes the combination of carbon dioxide and silicates. Carbonates and silicon dioxide (quartz) are formed if calcium is the metal involved (as it usually is); other metals can also serve to absorb carbon dioxide. Such reactions, Hart noted, take place much more rapidly in the presence of water, such as in oceans. On Earth at present, the limestone beds are in equilibrium with the carbon dioxide levels of the atmosphere; some new deposits are forming while some old deposits are dissolving.

Hart simulated the presence of life based on some upper limits to the mass of living organisms which Earth could support, on the chemical product of photosynthesis (less than total recylcing was called for, since deposits of coal and oil were laid down), and on various chemical reactions and the dissolving of gases in water.

The energy budget was simulated based on astrophysical theories about changes in the Sun's brightness, coupled with a complex model of the reflectivity or albedo of Earth's surface and atmosphere, and of the "greenhouse effect" which trapped excess heat.

His mathematical model of a planet seemed to behave as planets behave. Starting from reasonable initial conditions, it eventually reached conditions known to be in effect today. Along the way, Hart found some other surprising results—Earth may be a fluke after all.

What Hart found was that Earth seems to have balanced precariously between two ecological precipices: the "runaway greenhouse" syndrome which bakes a planet into a sterile sister of Venus, and the "runaway glaciation" syndrome which freezes the planet's oceans, covers them with dry dust, and presents an outward face barely distinguishable from Mars.

Hart's initial conditions called for an atmosphere of 84% water vapor, 14% carbon dioxide, 1% methane, and traces of nitrogen, ammonia, and argon. (This material could have outgassed in entirety, although Hart admitted a large portion of it could have been imported, relieving Earth of the necessity of outgassing those particular fractions.) An unbroken cloud bank covered the infant Earth, while atmospheric pressures rose and water condensed to form oceans (or, as today, *one single* ocean).

The amount of carbon dioxide decreased once the oceans were formed, through the mechanism of dissolving in the water and being precipitated out in the form of carbonate rocks such as limestone. That is where the vast majority of Earth's carbon dioxide is safely locked away today. As life formed and flourished, heavy hydrocarbons were built up in the oceans, perhaps developing like an oil slick many meters thick. Biological activity helped turn Earth's atmosphere into one consisting almost entirely of methane and other volatile hydrocarbons.

The effects of the steady production of free oxygen began to be felt (and the oxygen could only accumulate as rapidly as the carbon was stored away safely) when, more than three billion years ago, the continuous cloud cover broke up. Surface temperatures fell sharply; the atmospheric pressure fell to about half what it is today. By 2500 million years ago, the early hydrocarbon components of the atmosphere had been consumed by oxygen, leaving only nitrogen.

More than 10% of Earth's surface was covered with the planet's first glaciers, and Earth approached a climatic crisis. According to Hart's computers, the temperature had dropped 38°C; had Earth's temperature dropped only 2°C more (say, if Earth had been 2 million kilometers farther from the Sun), the glaciers would have continued to expand, reflecting more and more incoming sunlight, and causing the planet to plunge into a perpetual ice age in which the oceans themselves would freeze solid. "In every such case the computation was continued," Hart wrote, "often for an additional two billion years or even longer. . . . In not a single such case was runaway glaciation reversed."

A similar crisis had passed earlier in Earth's history when the temperatures had been high. Had Earth been only 8 million kilometers closer to the Sun, no oceans would have condensed; instead, the water vapor and increasing amounts of outgassed carbon dioxide would have surrounded Earth with a suffocating "runaway greenhouse effect" which would have maintained temperatures above 300°C. Life could never have formed.

The Gaia Hypothesis

The notion, based on Hart's computer modeling, that Earth at least twice barely escaped death—either the heat death of "runaway greenhouse" or the ice death of "runaway glaciation"—is frightening. In addition, if Earth is such a fluke, this has serious implications for the search for other Earth-like worlds in the galaxy.

There are those who disagree with this "blind luck" model of Earth evolution. The best formulation of the opposite notion is probably James Lovelock's Gaia Hypothesis, named for the Greek goddess of the world.[6] Lovelock suggests that Earth's biosphere has modulated the whole biosphere in order to maintain a habitable planet. Wild variations are thus prevented by some sort of feedback control over the environment; the entire Earth's biosphere can be said to be engaged in deliberate planetary engineering.

One of the most interesting tests for this theory deals with the problem of the variation of solar output, which astrophysical theory requires to have increased approximately 30% since the birth of the Solar System. Yet the geologic record indicates that the temperature of the biosphere has not varied widely at all. Hart's mechanistic model must invoke coincidental changes in the composition of the atmosphere, the timing of which is crucial; statistically, it's highly unlikely. According to the Gaia Hypothesis, however, Earth biology modulated the amount of carbon dioxide in the atmosphere, thus controlling the effectiveness of the "greenhouse effect" and consequently producing just the appropriate temperatures for optimal survival.

This hypothesis allowed Lovelock to predict that life would not be found on Mars. "There is no sparse life," he wrote prior to the Viking landings. "A planet is either living or not, just as a person is either living or a corpse." The existence of life completely remodels a planet and can survive surprisingly wide variations in external energy inputs through the application of some still mysterious control system. That did not happen on Mars, so life could not exist on Mars—so states the Gaia Hypothesis.

Still, without additional statistics, it is illogical to argue that "since we are here safe, we were inevitable." Obviously, if a thousand infant Earth-like worlds were formed throughout the universe, and over the eons all but one fell victim to the vagaries of climatic disaster, the inhabitants of that one surviving Earth (ignorant of the demise of siblings) might indeed pride themselves on their own inevitability and invulnerability, not realizing that Earth really *was* the fluke and that dead Venus-Mars-Moon-worlds were the norm. We just do not know.

The Role of the Moon

There's another factor in the evolution of Earth's biosphere, and in the prevention of the environment from swinging too far out of line leading to irreversible "runaway glaciation" or "greenhouse." That factor, ignored by Hart's model and by most other planetary scientists, is the effect of the presence of the nearby Moon.

As summarized in *Science* in an article by Alan Hammond[7]:

> Only the earth of the inner planets has a massive satellite, which may account for the planet's relative climatic stability. . . . On Mars, large oscillations in the obliquity or tilt of the planet's axis are thought by some to lead to gross changes in the Martian climate. The oscillations are due to an interaction between two dynamic phenomena—the precession of the equinox as the tilted axis describes a conical motion and the precession of the planet's orbital plane as the entire orbit wobbles in and out of alignment with the rest of the Solar System.
>
> The earth's obliquity changes very little, at present, because the presence of the moon shortens the equinoctal precession period,

precluding a resonant interaction with the orbit plane precession. Without the moon, however, the earth's obliquity would oscillate even more than that of Mars, leading to far greater climatic instability than we presently experience and endangering the course of biological evolution.

We'll examine these factors again in a subsequent chapter, when we consider natural ways in which the climate changes. But the lesson is clear: the presence of the Moon's guiding and stabilizing gravity may well have prevented Earth from slipping over the edge of climatic disaster during the billions of years of its history. And since the existence of such a relatively large moon appears to be a statistically unlikely occurrence, most other proto-Earths in the universe may not likewise have been so fortuitously saved. Our Earth looks, then, more and more unique—and therefore precious.

Natural Climate Changes

The climate of Earth already varies under many cycles and under many influences even without human interference. There has been growing awareness of the wide climatic fluctuations of the past (the ice ages are only an extreme example), and of the multitude of varying factors which together drive Earth's climate. Only recently has human activity become a significant climatic factor on a regional or perhaps even a global scale; the true extent of human effects on Earth is still a subject for debate. Waiting in the wings are plans and suggestions for future projects which will deliberately alter the face and the climate of Earth.

Earth's climate has been varying over the eons, but our presence here today is proof that the average temperature never got above the boiling point or below the freezing point of water; there has always been some liquid water on Earth. This may have been only an accident, or it may have been the result of feedback from the actions of the biosphere (the Gaia Hypothesis). But whatever the mechanics of these fluctuations, they have been in motion since long before human activities began to significantly make a dent in the planet's ecology.

Following a period of very benign climate several thousand years ago, Europe endured a decrease in rainfall and a drop in the average temperature. Occasional long droughts were interspersed with relatively brief periods of increased rainfall. These dry periods occurred from 2200-1900 BC, from 1200-1000 BC, and from 700-500 BC. The climate then

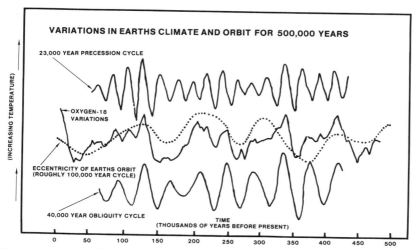

Varying factors which influence climate are shown here. Over the short run (top), the variations in Earth's orbit and the resulting changes in sunlight intensity and distribution seem to have the largest effects. Over the long run (bottom), numerous influences—both external and internal—can create vast climatic changes. Humanity may someday be able to duplicate those influences and replicate their results by changing planetary climates.

became wetter and colder, with the resultant spread of peat killing off forests in many areas. Lakes in the Alpine region rose, destroying the stilt homes of many societies. Then, within 500 years, rains again decreased and the temperature rose to modern levels.

In the mid-fourteenth century, the climate again turned cold, during a period now known as the "Little Ice Age." This period reached its worst in the 1600's, coinciding with a recently recognized absence of sunspots (the "Maunder Minimum") and *polar aurorae* (the "northern lights," a result of storms on the Sun). Judging from notebooks of astronomers, the Sun rotated differently than it does today, with a greater differential rate between the equator and higher solar latitudes.

This was the reign of Louis XIV, the self-styled "Sun King." Indeed, during this period, the Sun's face was remarkably unblemished, but the Sun's effect on Earth caused climatic catastrophe.

The Viking settlements in Greenland were wiped out. The Thames River repeatedly froze over (eight times in the seventeenth century), as did New York harbor. The plant virus which caused the great Irish potato famine flourished in a climate so different from today's that the virus cannot now even grow in Ireland.

A brief warming trend in the early twentieth century was noticed on instruments, giving an average increase of 0.2°C from 1899 to 1939. Yet this average did not reflect some even greater changes which were occurring locally: some areas in the United States rose 1°C, parts of Scandinavia and Mexico rose 2°C, and Spitsbergen was a full 5°C warmer.

In recent decades, a cooling trend has set in. Increased snow cover has been noted on Baffin Island, and there is now pack ice around Iceland after more than half a century's absence. Glaciers in the Alps have begun advancing again. In the winter of 1978–1979, four of the five Great Lakes were completely covered with ice—the first time this has happened since record-keeping began.

The biosphere has taken note of this change. Britain's annual growing season shrank by ten to twelve days between 1951 and 1967. In the United States, mild climate creatures such as armadillos, which appeared as far north as Nebraska during the warming period, have headed back south again.

Fractional changes in degrees may not seem like much, but averaged over an entire season they can determine the gross climate of an entire region. A fall of a single degree centigrade, for example, might seriously endanger the entire Canadian wheat crop.

The Ice Ages

These variations, of course, pale in comparison with the great Ice Ages (and by geologic standards we are living in a "brief" interglacial period of the present Ice Age). Understanding the causes of such small changes will be much more difficult than discovering the origins of the large freeze-ups. This in turn makes it harder to determine what environment variables might be subject to artificial manipulation in order to forestall the reoccurrence of such glacial epochs.

Over 20,000 years ago, nearly half of Earth's land surface was covered with ice to a depth of a kilometer or more (up to a mile thick). Other ice ages have been detected in geologic records 550 million years old (the Cambrian period) and in the late Paleozoic period 250 to 300 million years ago. Polar ice then seemed to have vanished until it reformed on Antarctica about five million years ago and on Iceland and Greenland about two and a half million years ago.

Near the maximum of each of these glaciations there occurred periods of extremely arid worldwide climates, perhaps caused by the lower surface temperatures of the oceans which led to a much lower evaporation

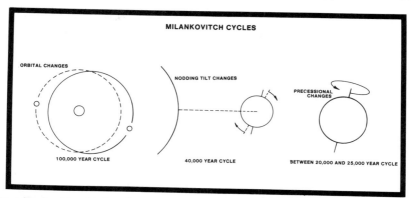

The Milankovich orbital cycles are considered to be a valid explanation for the occurrence of ice ages on Earth. They also effect the climates of other planets as well.

rate. During this interlude, there were widespread deposits of finely divided surface dust (the *loess* dust) laid down; these thick dust beds are today one of the main sources of particles in the atmosphere, with all the resultant effects of such dust. Indeed, perhaps air-borne dust deposits falling on the glaciers was a factor in speeding their melting by decreasing their albedo.

Dust plays a role in another glacial theory from British astronomer W.H. McCrea, who has noted that Earth has just recently left the "dust lane" bordering the Orion Arm of the Milky Way galaxy; the action of dust in space may have blocked sunlight and cooled the Earth; (although other scientists postulate that additional dust would have stoked the fires of the Sun, resulting in a net *increase* of solar radiation reaching Earth).

Whatever the actual mechanisms, terraformers should note them and should be prepared to mimic them for planetary engineering projects. Dust, after all, is a plentiful resource in the Solar System, or can be easily manufactured from passing asteroids.

The most widely accepted theory for explaining the ice ages is credited to a Yugoslav scientist named Milankovich,[9] who traces Earth's climatic variances to the different levels and distribution of incoming sunlight caused by periodic oscillations in Earth's orbit and pole. The irregularity (or *eccentricity*) of the orbit varies over about 90,000 years due to the gravity of other planets: the polar tilt angle varies from 22° to 25° over a period of about 42,000 years; the polar tilt direction *precesses* or wobbles around the sky in a period of about 26,000 years, due to the effects of the Sun's gravity on Earth's equatorial bulge.

80   New Earths

Resulting changes in heat distribution seem to be able to account for climatic changes occurring on the surface, particularly the Ice Ages. The lesson is that fairly minor variations in incoming sunlight can have far-reaching effects on the planet's climate. There are several artificial techniques by which incoming sunlight can be modulated—hence climate can be controlled.

The Human Role in World Climate Change

All of these changes were taking place without regard to human activities. Changes taking place today are on the same scale as the natural changes, so human activity, however influential on a local or regional level, may not be nearly as disruptive as some environmentalists warn. One observer has cynically remarked that the need of some people to blame contemporary climatic changes on technology is "masochism of the over-affluent," or delusions of ecological grandeur.

"It will be difficult," noted MIT's 1971 study, "to identify any man-made effect because, first, with our present state of knowledge, we do not know how to relate cause and effect in such a complex system, and,

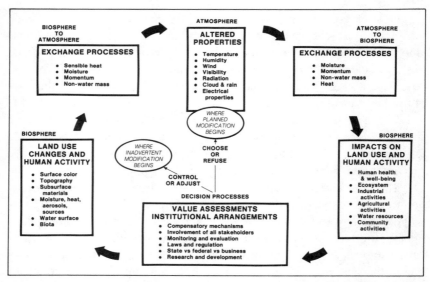

Human interference in "natural" climatic variations is beginning to have measurable effect, and calls are now being made for deliberate climate engineering. This flowchart from a meteorology conference in 1977 shows the various interrelationships and consequences of human activities and natural forces.

second, manmade effects will be obscured by the natural changes that we know must be occurring."[10]

Climatologist Stephen H. Schneider, warning the world to avoid climatic disaster in *The Genesis Strategy* (1977), agrees: "The global climatic effects of human activities cannot yet be proved larger than those natural forces responsible for worldwide climatic variability over the past few thousand years. . . ."

A comparison of available energies demonstrates this problem. The net solar radiation is about 100 watts per square meter on Earth's surface. Some urban areas give off about 10% of that value (and by the next century there will probably be wide industrial areas where the manmade heat will be equal in strength to the Sun's heat); but on a world-wide basis, humanly released energy comes to about 0.015 watts per square meter, several times less than the very weak natural heat flux from the interior of Earth. That global average may climb to 0.5 watts per square meter by 2100 A.D., at which point it could be of planetary significance. Yet the Sun's energy seems to naturally vary far more wildly than this human contribution.

Natural processes release tremendous energies of their own. Using scales designed to measure nuclear bombs (small A-bombs pack several tens of kilotons of TNT equivalent, while large H-bombs are measured in the tens of megatons), nature is blasting Earth continuously with forces far greater than those in today's nuclear arsenals: tornado funnels can give off up to 50 kilotons; a thunderstorm tower is the equivalent of 500 kilotons; a moderate Atlantic depression may release a thousand megatons; a large Atlantic depression would be gauged in tens of thousands of megatons.

Manmade Deserts

We have already recognized ways in which human activities have grossly altered regional climates, but these projects have taken centuries and have usually been inadvertent—and locally disastrous. The primary mechanism is in the deforestation of vast areas caused by the domestication of animals which feed on tree seedlings.

Parts of Africa and south-west Asia have been transformed into semi-desert in historical times, probably due to this mechanism. Dense forests in the mountains of Turkey, Iran, and Afghanistan, plus regions along the Mediterranean, Europe, the eastern United States, and Japan have van-

Photographs from space show how human activity has markedly altered the albedo of Earth's surface on a local and regional scale, thus creating new patterns of rainfall, dust formation, and winds. The Egyptian-Israeli border (arrow) clearly shows on this Gemini photo, due to overgrazing by Bedouin tribal goats on the Sinai Desert.

ished, in part cut down for farm land and for building material, in part the victim of the domestic herds. The Rajasthan desert in India is also probably manmade via this process.

Goats owned by Bedouin tribes in the Sinai have defoliated enough land on the pre–1967 (and post–1979) side of the Egyptian-Israeli border to make that political boundary clearly visible from space because of the albedo difference. Similarly, a Skylab photo of a national park in New Zealand clearly shows that the non-grazed areas inside the park fence

How Earth "Happened"   83

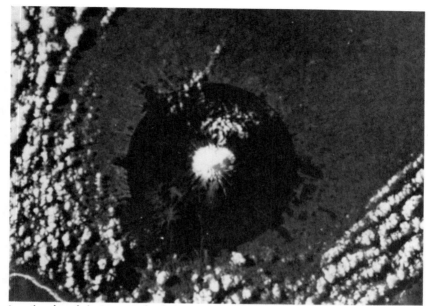

A national park in New Zealand shows up as a dark almost perfect circle because a boundary fence keeps out grazing animals. These same activities probably deforested major portions of the Mediterranean and Indus regions four or five thousand years ago, creating new deserts and destroying civilizations.

have denser vegetation than outside. The same thing was happening in the Sahel region of West Africa prior to the recent tragic droughts. Similar deforestation is happening in South America, Africa, and south Asia.

Destruction of jungles such as the Amazonia is irreversible. Soil moisture is no longer maintained; the increased rain impact velocity on the ground increases runoff and erosion while taking advantage of a lessened soil cohesiveness. The micro-climate of the surface is changed, with a higher surface wind speed decreasing the growing season and increasing the dust production rate. There are also frequent flash floods interspersed with dust storms.

Almost all of the savannah grasslands of the tropics are considered to have been manmade; primitive slash-and-burn agricultural methods would have changed the soil acidity to the point where new types of plants could grow there. This probably led to the establishment of the pine forests in the southern United States. Perhaps as much as 20% of Earth's total continental land area has been drastically changed by human agricultural activities of this type.

Trees are useful for more than just albedo adjustments and shade, so human deforestation activities have additional, subtle side effects. Trees produce oxygen from carbon dioxide, and they also transport water from below the ground surface into the air. This process is called *evapotranspiration*.

A single apple tree may move 9,000 liters (2,000 gallons) of water into the air in a six-month growing season. This factor is so important that there are cases in which the cutting down of a forest has resulted in the region turning into a swamp because the flow of ground water that had previously been dispersed by the trees is no longer being conducted. This local effect does not include the more widespread effects of the lowered humidity downwind of the former forest.

These problems cause anxiety when they are considered in conjunction with the statistics of worldwide land use. Of Earth's land area, 25% is potentially arable, although little more than 10% is now actually in agricultural use. Even excluding the humid tropics and regions without plausible irrigation sources, we could still more than double the crop area of the world. This would only increase the planetary albedo by 0.2%, but it would have much more serious local effects, and those effects may ripple across continents before we know it—climatic dominoes.

This examination of Earth's evolution, and the directions in which that evolution is proceeding even today, has set the stage for the next step in our study: the deliberate intervention of human activity at preselected fulcrums in Earth's biosphere so as to alter climatic features of the planet. We are now ready to contemplate basic planetary engineering, with Earth as the first subject.

Give me a place to stand, and I will move the Earth.
Archimedes

4

Restructuring Earth

Change, not stasis, is the natural order of this planet, and humankind will either guide and steward that change, or become the victim of it, along with other forms of life. Human civilization is too complex to tolerate the large climatic variations which have occurred frequently in the past and have resulted in the collapse of civilizations, the onset of famine and disease, and, not infrequently, the extinction of entire species. An examination of the ways in which such climate changes have occurred can provide terraformers with possible techniques to combat such future variations on Earth, preserving our planet and our civilization.

Earth's climate is now seen to be in a constant state of flux, fed by external forces (such as changed solar energy delivery) and by internal forces (such as the long period of time in which the oceans absorb and release vast amounts of energy, or the effects on the planetary albedo of the spread of new types of surface vegetation). These changes have been going on since Earth formed.

86 New Earths

Artist's concept of Shenandoah Valley after Earth has suffered a "runaway greenhouse effect," raising surface temperatures to 400° C.

Human activity has been affecting local climates, and may now be one significant factor in global climate change. These activities and their consequences are all accidental and generally result in unpleasant surprises.

We have reached the point where many scientists and engineers have proposed deliberate climatic modification efforts, on scales more vast than those of rain making or river damming. By means of effects already being accidentally produced by human civilization, would-be planetary engineers have dreamed up many schemes.

Some good reasons to conduct such projects (adapted from *The Genesis Strategy*,[1] might be to:

1. Offset an inadvertent climatic change due to previous human activity;

that is, to try to repair damage already blundered into by ignorance
2. Relieve drought on a regional or continental scale, all the while making sure not to steal the rain—it's called "cloud rustling"—from nations farther downwind
3. Relieve or prevent flooding; avoiding rainfall is only one way of doing this, since snow melting is another prime cause of spring floods and might be ameliorated by modulating incoming solar radiation to allow the snows to melt gradually rather than catastrophically
4. Increase fresh water supplies, both surface and sub-surface
5. Offset a gloomy long-range weather forecast; that is, interfere with the natural course of events which might be leading the world back to another "Little Ice Age" or in the opposite direction, toward ice cap melting
6. Stabilize climate; this would refer to countermeasures against climatic factors tending to drive a climate to uncharacteristic extremes of heat, cold, rain, or drought
7. Improve food production
8. Gain a military or economic advantage

Other conveniences belong on this list as well: improving surface and air transportation efficiency and safety; improving climates generally to please local populations; and controlling plant and animal pests.

Perhaps the most widespread effect of human activities on Earth is the gradual but seemingly inexorable increase in atmospheric carbon dioxide over the past century. From an estimated figure of 290 parts per million before the Industrial Revolution, the level has reached more than 320 parts per million and could be 400 parts per million by the end of this century. By that time, the additional carbon dioxide would be sufficient to increase global temperatures by a full degree, due to a stronger "greenhouse effect."

About half the carbon dioxide released by the burning of fossil fuels ends up contributing to this increased level; the other half, it has been assumed, is dissolved in the oceans or absorbed by land biota. Estimates have been made which claim that the burning of fossil fuels absorbs as much as 15% of all the biological production of oxygen.

And yet our true ignorance of even the simplest ecological cycles of Earth was brought home in 1978–1979 with a revolution in the concept of the carbon dioxide cycle. According to new results (which are not universally accepted by any means), the primary source of released carbon dioxide is not industrial activities but the steady deforestation of

South America and Africa. Carbon dioxide is being produced from the subforest humus when it is exposed to the air, and from the burning of the trees.

According to Dr. John Adams and his colleagues at the University of Sao Paulo in Brazil,[2] the clearing of about 100,000 square kilometers (40,000 square miles) of forest per year is both increasing the release of carbon dioxide, while at the same time decreasing the ability of the biosphere to reprocess that carbon dioxide. "Without concurrent deforestation in this century," the university team reported, "the build-up of atmospheric carbon dioxide would have been very minor at best."

A lesson to be learned from this controversy is that the interchange of carbon dioxide between the atmosphere and the ocean is a very poorly understood phenomenon. It turns out that several times as much carbon dioxide is being removed from the air as can be accounted for by current models of the ocean; yet there is nowhere else it can be going. On other planets, we expect to use open water as a mechanism for carbon dioxide concentration control, but we do not even know yet how it works here on Earth!

By the way, one interesting feedback mechanism probably still holds, and has significance for planetary engineering. The ocean's ability to absorb carbon dioxide is a function of temperature—the warmer the water, the less carbon dioxide can be absorbed. This is a *positive feedback system* (meaning it's unstable), since a carbon dioxide buildup increases planetary temperature, driving dissolved carbon dioxide out of the oceans, increasing the gas buildup and the resulting temperature rise, and on and on.

A similar gap in our knowledge exists with the carbon monoxide which human activity is pouring into the air. This gas only lingers about a year before being transformed into something else, but nobody has any good ideas about what acids or other materials in the atmosphere might remove it.

Other consequences of the burning of fossil fuels are easier to trace and, you guessed it, they are bad. The oxidation of coal, natural gas, and oil results in the injection of substantial amounts of sulfur dioxide into the air, where it mixes with water and creates sulfuric acid droplets (like the atmosphere on Venus, but not quite as strong).

When this substance precipitates as rain or snow, it has a very deleterious effect on the biosphere, particularly in Scandinavia, northeast United States, central Europe, and Britain. In southern Norway, for example, salmon has disappeared from thousands of forest lakes; forest growth is greatly reduced, due in part to excessive leaching of nutrients from the soil by the acid rain. A similar reduction in forest growth over the past twenty years in the northeastern United States may also be traced to the acid rains.

Some observers have questioned whether or not the sulfuric acid droplets are primarily caused by human activities at all since there are natural sources as well. Whatever the cause, the important thing is to study the effects of such trace contaminants on worldwide biological activity, in order to be able to model them on other planets.

Contaminants in general are a very sticky problem. If human production of carbon dioxide remains controversial, human production of heavy metal contamination is not. Mercury, copper, cadmium, lead, and other metals have had catastrophic impacts on the biosphere when their concentrations have been too great. The presence of excess concentrations of such poisons on other planets could be a critical factor in the artificial development of biospheres—unless one uses biota for which such contaminants are *not* poisonous.

Polar Caps

Planetologists call the frozen part of Earth the *cryosphere*. Ice covers 10% of Earth's land, up to three kilometers (two miles) deep in places; another 10% is in perma-frost, where the layers just below the surface (which thaws in the summer) can be frozen to depths as great as half a kilometer (about 2,000 feet).

There seems to be some sort of balance mechanism maintaining that ice. Geologists believe that there has always been *some* surface ice, but not too much, since Earth has never completely frozen over. At the height of the last ice age, ice covered about three times as much land as now; at other times, ice was rare and restricted to polar highlands.

Ice stays cold by resisting the input of solar heat; the mechanism, as already described, is a high albedo. Even in the summer months, the radiation budget of the ground atmosphere system is "in the red" in the polar regions. *Advective heating,* the transport of warm air from the equatorial regions, compensates for this deficit.

The poles are also deserts, judging from precipitation. Very little moisture falls, but that which does, stays around a long time.

The geography of Earth's polar regions is also peculiar, leading to a situation which may be more extreme than otherwise should have occurred: the south pole is centered on a continent surrounded by ocean; the north pole is located within a deep ocean basin which is nearly completely landlocked. Both of these conditions may tend to increase the stability of the ice sheets from outside perturbations.

Still, there is some concern about the stability of the Arctic ice pack. Small changes in heat input (which includes albedo variation) might cause very large variations in ice area, and strong variations in temperature at middle latitudes. Recall that the relative heat imbalance is a major factor in driving the poleward air circulation. Any changes in this amount of imbalance would likely have major effects on the weather patterns of the whole planet.

On colder planets, polar regions will be even more crucial, since the right combination of orbit and planetary axis could plunge a polar region to temperatures cold enough for much of the atmosphere to freeze out as snow. Many specialists used to believe that this had happened on Mars, and several fascinating terraforming schemes were developed to counteract this effect.

Thus, another fulcrum of Earth's biosphere at which artificial levers may be placed is the arctic ice pack. It seems to maintain a precarious existence, while having an influence on world climate out of all proportion to its size and fragility.

The ice itself is about three meters thick in late winter, melting down to about two meters thick by the end of summer. Sixty percent of the ice near the pole is five years old or more; two percent may be up to twenty years old. The area of the floating pack increases by 20–25% through the winter, and can cover as much as 90% of the arctic ocean.

When present (and only 5,000 years ago, the ocean was apparently ice free), the floating ice allows the air to cool more intensely in winter and warm more intensely in summer, since the ice serves as an insulator between the air and the water. The temperatures just above the ice in winter average about $-30°C$, but only $-2°C$ in water just below it.

The equator-to-pole temperature gradient is one of the most important driving forces on planetary weather patterns. The winter semipermanent polar anticyclones provide one of the anchors for the global atmospheric circulation; with a decrease of the equator-to-pole temperature difference, the subtropical anticyclone belt would be displaced northwards, drastically redistributing Earth's desert and rainfall zones.

The disappearance of this ice by deliberate or accidental human activity would have immediate local consequences as well. There would be a more moderate coastal climate around the Arctic Ocean, but precipitation would increase substantially, perhaps leading to renewed continental glaciation. While some regions of permafrost might be opened for cattle pasturage, others could spawn new glaciers.

An ice-free Arctic with its milder climate might contribute to the melting of the Greenland ice cap, however, and that could cause trouble. The Greenland ice is in an unstable situation: the snow-fed glaciers are at high altitudes in contact with cold air; as the ice melts, the elevation of the surface decreases, bringing it into contact with denser, warmer air which accelerates the melting. The melting of most of the Greenland cap would raise world sea levels by about 7 meters (about 23 feet).

On the other hand, the increased snowfall over Greenland might counteract mid-summer melting and preserve, even extend, the present cap. Clearly, more study of this system is needed.

Warming of the arctic ice from ocean currents (mainly from the North Atlantic) and from increased solar heat are natural ways that the ice can be melted. There are, however, numerous side effects of human activities which could also melt the ice: increased global temperature due to strengthened "greenhouse effect"; increased salinity of the Arctic due to the cutoff of fresh water from the Siberian rivers (resulting in a lower freezing point for the water); deliberate spreading of dark material over the ice to absorb more sunlight; oceanic manipulations, perhaps with nuclear bombs, to mix warmer, saltier deep waters with the surface; the construction of a dam across the Bering Strait.

This melting would cause the most severe climatic changes since the last ice age maximum, according to projections of climatologists (although, since the ice is floating, it would *not* raise sea levels). Snow squalls and increased over-land precipitation would result. This would then increase regional cloudiness, with an unknown net effect on albedo; the clouds would tend to prevent ground heat from escaping into space at night, reducing continental cooling. The new increased snow cover around the Arctic basin would insulate the ice below. Air temperatures could be 20°C warmer in summer, but only 5°C warmer in winter; the net temperature increase south of 50° latitude would probably be about 1°—significant but not overwhelming. The displacement of rain patterns could, however, be catastrophic for the entire northern hemisphere.

There are (as listed above) several methods by which civilization could tamper with the amount of absorbed solar energy over the Arctic

REGIMES OF CLIMATIC STABILITY

Climates may exist in different modes of stability, illustrated here by the position of a marble on a surface. In the unstable mode, even the slightest push in any direction will induce larger and larger off-center forces ("positive feedback") which propel the marble away from the initial conditions. A situation may be "interval stable" if the marble has a small region within which it can wander without reaching the "knee" of the downward-sloping curve. The marble's position is stable if off-center motion induces forces propelling the marble back to its initial condition ("negative feedback"). Lastly, conditions might be such that a sufficiently powerful push might propel the marble up over a hilltop and down into another stable valley; this could be considered "dual stability."

regions, but one in particular is worth singling out: the spreading of dark soot over ice. This idea has been repeated so frequently that it sounds like a metaphor. But it has already been done, first in Chile in 1969 to help melt some of the Coton glacier in the Andes, and elsewhere since then.[3] This same method might produce different results over deserts, where blacktop paved areas would result in massive hot air updrafts, drawing in moist air from nearby oceans and compelling it to condense into rain (as long as it doesn't wash off the black paint!).

The British Meteorological Office in Bracknell, England has devised a new computer model of Earth's atmosphere. When tested, it seemed to produce climatic patterns similar to those observed in nature, and many cause-and-effect sequences confirmed predictions made with less sophisticated models. But there were some surprises, as reported in 1979 by Dr. John Mason.

"We assumed the Arctic sea ice melted," wrote the director general of the Meteorological Office, "being replaced by a water surface. . . . The major effects of removing the ice, apart from the obvious one of warming the polar regions, were to weaken the intense polar anticyclone (high pressure system), to diminish the strength of the middle-latitude westerlies and their associated depressions (low pressure regions), and to produce a significant cooling of up to 8°C in middle latitudes, especially over the United States, eastern Siberia, and western Europe. The last result is rather unexpected and serves to illustrate the limitations of intuitive judgments in dealing with such highly interactive, non-linear systems."[4]

The model was also applied to the question of desertification in the Sahara, and to ways of counteracting such climatic trends. Reported Mason, "We made comparisons between one set of simulations in which the Sahara desert, placed between moist zones representing the North African coastal strip and the savanna region to the south, was made initially dry with no soil moisture, and another series in which the Sahara region was made initially wet with 10 centimeters of soil moisture. In the first series, shallow depressions crossed the region but produced little precipitation because there was no surface moisture to feed and maintain them. In the second, the surface temperature over the wet ground fell by as much as 20°C, cooling extended up to heights of 5 kilometers (18,000 feet), and major depressions developed, producing widespread rain, heavy in places, which persisted during the twenty days simulated in the experiment. It seems that once an area of this size becomes wet it tends to stay wet. . . ."

The important point Mason is stressing in the Sahara example is that nature often provides a region (or a planet) with two widely separated stable regimes, each of which will remain in its own state indefinitely unless perturbed into the other state by an external force—either natural or artificial. This is a major opportunity for planetary engineering, since a changed climate need not require *continued* application of the changing force if we can design it right. Mars, as we'll see shortly, just may be such a "two-state" situation, to be changed from the present unhealthy one to a far more hospitable one—which will subsequently *stay* that way!

Global Engineering

For many engineers, the term *planetary engineering* implies great architectural structures rather than climatic alterations (although that may result from the construction of such a project). A century ago, George P. Marsh wrote that "there is little in the way of mechanical achievement which seems hopelessly impossible, and it is hard to restrain the imagination from wandering forward a couple of generations to an epoch when our descendants shall have advanced . . . far beyond us in physical conquest," even to the point of leveling the Alps for agricultural purposes. Fifty years later, R. L. Sherlock wrote *Man as A Geological Agent,* claiming that human power to transform the face of the planet would remain immense.[5]

As quoted by geographer Richard B. Cathcart in a paper submitted to (but not presented at) the Houston terraforming colloquium, Glenn T. Seaborg of the old Atomic Energy Commission defined terms by saying, "Planetary engineering includes weather modification, watershed control, and all large-scale Earth modification activities." Geophysicist Fred Singer, while with the U.S. Interior Department, told an audience in 1963 that "we are moving into the age of planetary engineering and now is the time to discuss proposals for modifications and their consequences before the need of such modifications becomes urgent." The need for such modifications has already arrived as many engineers believe who wistfully consider planetary engineering projects such as those put forth in Willy Ley's book, *Engineer's Dreams,*[6] which proposes the wholesale restructuring of our home planet.

For most specialists, the planetary engineering of the immediate future still remains primarily in the field of water control—irrigation, dams, canals, and so forth.

At present, there are about 300 manmade lakes and reservoirs on Earth larger than 100 square kilometers (40 square miles); the total manmade water surface is about 300,000 square kilometers (120,000 square miles). Proposals made over the past few decades could multiply that by several dozen times.

There is, for example, the long-standing suggestion by Robert Panero to dam the Amazon, a river which contains one-fifth of all the fresh water in the world (ten of its tributaries are *each* larger than the Mississippi). Since the slope of the region is so gentle, the dams need not be very high; an excellent location for the first dam would be near Obidos, about 1600 kilometers (1000 miles) from the coast, where hills flank the river. The resulting lake would be only 30 meters (100 feet) above sea level, but could become as large as 300,000 square kilometers (120,000 square miles), equal in area to all previous manmade lakes combined. The dam could cost $1 billion, but would provide electrical power for a major aluminum smelting industry.

Plans originated by Herman Sorgel fifty years ago to create a "Chad Sea" and a "Congo Sea" in Africa through the construction of appropriate dams could put up to 10% of the African continent under water, providing irrigation, transportation, and electricity—as well as ecological upheavals of unpredictable severity. As long ago as 1902, a Melbourne University professor named J. W. Gregory called for the flooding of Australia's Lake Eyre, a dry, below-sea-level depression inland of the fertile coast. And the Qattara Depression in Egypt has often been named as a candidate for flooding, in order to irrigate vast areas of the Sahara and provide electrical power as well.

The Qattara project deserves more attention from would-be terraformers because in many ways it could be an analog (a dry run, but with water) of the flooding of Mars. The supersaline soil of the eastern Sahara is, if anything, more hospitable than the Martian soil; but the techniques involved with creating a new ecology along the shores of a new body of water will have much to teach planners with their eyes on Mars.

The proposed Bering Strait Dam also needs to be mentioned, especially in light of what we now know about the sensitivities of the Arctic Ocean and the possible consequences of upsetting the thermal balance. The Bering Strait is only 75 kilometers (45 miles) across and, at most, 60 meters (190 feet) deep, forming a very shallow passageway between Asia and America through which very little water presently flows. But it is enough: cutting the flow entirely might alleviate the harsh east Siberia winters, but it could also divert Arctic currents into the icy Labrador current past Greenland, with consequences for northeastern Canada and, through the effect on the Gulf Stream, all of northern Europe. The chain of cause-and-effect on the planetary climate machine is extensive indeed—as we are just beginning to realize!

There are some unappreciated drawbacks of such grandiose proposals. The first concern is in the redistribution of weight over the surface. In the case of the Amazon dam system, more than 170 billion tons of water would be gathered together. Such new stresses cause earthquakes: in 1963 a 6.1 Richter scale quake hit the shores of the artificial Lake Kariba in Rhodesia; in 1967, 200 people were killed in India near the Koyna Dam southeast of Bombay—and that reservoir only involved two billion tons of water, but it set off a series of sharper and sharper quakes, reaching 6.4 on the Richter scale.

That's a foretaste of what superdams could cause elsewhere on Earth, and what massive relocation of water and other volatiles could do on other planets.

Russia's Superirrigation Plans

Probably the most ambitious planetary engineering project with any real chance of being attempted in the next few decades is the Soviet scheme to redistribute the waters of Siberia. Most of the large rivers flow into the Arctic Ocean, while vast regions south of them, in Kazakhstan and nearby areas, are desert, badly in need of water. The land-locked Caspian Sea (misnamed since by any standard of geography it's a lake) is drying up due to the diversion of most of the Volga River into irrigation projects; already the Caspian is more than 100 feet below sea level, and, deeply injured by vast industrial and petrochemical pollution, is becoming a new "dead sea." Planetary engineering is being proposed as a way to repair the disastrous manmade climatic results of inadvertent human actions.

The most spectacular part of such a system would be a dam at the confluence of the Ob and Irtysh Rivers; the earthen structure would be 75 meters (250 feet) high and at least 65 kilometers (40 miles) long. The waters backed up in the new reservoir would flood an area the size of Great Britain, an area that is now impassable and economically useless marshland. Water would back up the Irtysh and Tobol Rivers, and thence would be conducted by a canal 900 kilometers (575 miles) into the Caspian Basin. Another approach is to dam the Pechora River west of the Urals and carry it by canal to the Volga, and thence again to the Caspian Basin.

Climatic effects would be inevitable, as even the proponents of the project admit. "It is important to emphasize," wrote Dr. Mark L'vovich, head of the hydrology department of the Soviet Academy of Science's Institute of Geography, "that some negative consequences are bound to arise from any such drastic transformations of the environment—the problem is to minimize them."[7] Such potential eco-catastrophes include alteration of the Siberian high air mass which has a powerful influence on Asian weather to the east and southeast, in particular on the winds that set off the Chinese monsoons. And the reduction in fresh water inflow into the Arctic Ocean would increase the water's salinity and lower its freezing point, threatening to destroy the fragile ice cover—with all the implications which we have previously examined.

Soviet hydrologists have already been caught off guard by unforeseen consequences of earlier irrigation projects, L'vovich admitted. "One major problem has been the formation of widespread shallow waters and marshes along the edges of the reservoirs. Another has been the reduced water quality, the growth of weeds and the appearance of water eutrophication, choking on an overabundance of algae. . . . We did not pay enough attention to fish resources at the beginning. The sturgeon nearly died out. . . ." And Dr. A.G. Babayev, director of the Deserts Institute of the Turkmen Academy of Sciences, pointed out another unpleasant aspect of terraforming: as biological tools complete their tasks successfully, they would create conditions beneficial to new species but fatal to themselves. In Babayev's words, "If the desert were to receive abundant water, everything that grows in it now would die." The biosphere will be completely overturned, a process of *planned extinction* which will be the norm for planetary engineering.

Whatever the objections and cautionary warnings, the river reversal project seems certain to move ahead. In its massive scale, it appeals to the Russian spirit. The project has been discussed for generations, since the pre–Revolutionary engineer Demchenko first suggested it. Modern techniques now make it feasible, as demonstrated by a canal-digging test blast of three small atomic bombs set off in a row in 1971. The resulting crater was 700 meters (2300 feet) long, 330 meters (1100 feet) wide, and up to 15 meters (50 feet) deep—and all radiation was reportedly safely contained. The way seems open to the project, whatever the ecological warnings.

And that could be an inkling of future problems in regulating climate modification schemes. We may talk as much as we like about the absolute needs to understand all consequences before acting, but we still are face to face with the reality that such caution has never been the rule in the past, and is unlikely to be so in the foreseeable futur
nations may be forced into hasty modification projects o
A 1974 CIA study on climate control made just that
likely, perhaps, would be ill-conceived efforts to undertake drastic cures (where climate change caused grave shortages of food) which might be worse than the disease; e.g., efforts to change the climate by trying to melt the arctic ice cap."[8] The CIA's poorly disguised accusation against Soviet climatic engineers is not just a knee jerk dig at Moscow: the Soviets *do* have problems with their agriculture; they *do* have access to the arctic ice; they *do* have a history of grandiose (but ultimately disastrous) engineering and technological programs; and they *do* have a tradition of ignoring protests from countries beyond their borders. It is not a pleasant prospect.

A Terraformed Nation Today

For all the proposals to create new bodies of water through dams, there are also proposals to create new stretches of dry land through drainage canals and dikes. Draining swamps has been carried out by engineers for literally thousands of years. At the opposite end of history lie the suggestions to drain the entire Mediterranean, the North Sea, the Gulf of Mexico, or the Sea of Japan.

Between these extremes lies an excellent case study in small-scale terraforming: Holland. Here is a nation where half the population lives on land stolen from the North Sea, where only vigilant technology prevents a catastrophic reversion to the natural state. Even such a mentality cannot block occasional natural disasters, such as the storms in the early 1950's which killed 10,000 Hollanders.

Yet life goes on for the nation, repairing the ravages of nature and pushing the sea back step-by-step to claim new dry land. New ecologies must be established. New communities must be formed and financed. New generations must be trained in vigilance against an adversary more dangerous than any faced in war: nature itself.

Such a world view should find the concept and the implementation of planetary engineering "natural." Would-be terraformers have much to learn from the history, sociology, and psychology of the Dutch. They have a proverb: "God made the world, but the Dutch made Holland."

Hurricane Steering

Techniques of artificial energy direction via space mirrors and artificial evaporation reduction are applicable to the control of the mightiest energy systems on Earth—hurricanes. Combined with cloud seeding to induce rain (and to induce condensation which thus releases tremendous amounts of energy stored in the water vapor), these planetary engineering techniques suggest that hurricane control is an achievable goal.

During "Project Stormfury," meteorologists experimented with manipulation of hurricane forces and directions. "Stormfury" tampered with many hurricanes in a ten-year period—Esther in September 1961, Beulah in August 1963, Debbie in August 1969, and Ginger in September 1971—and positive results appear to have been achieved. That is, things happened to the hurricanes which probably would not have happened without seeding. But the natural factors which govern the hurricane's course are still only vaguely understood.

Restructuring Earth 99

Hurricanes may be steered away from vulnerable coastlines by the use of space mirrors and dust injection. Such an imminent possibility raises environmental issues which also apply to terraforming.

The dangers of such experimentation are obvious, as are the potential benefits. On the balance, most observers think it is worth it, but warnings such as this one, from E.C. Barrett's *Climatology from Satellites* (1974), must always be kept in mind: "Clearly, hurricanes must transport enormous quantities of heat towards the poles, thereby contributing not a little to the heat balance of the Earth/atmosphere system. There are good reasons for arguing, therefore, that it would be premature at this moment to contemplate frequent, large-scale hurricane control; potentially enormous and complex climatic consequences . . . might be expected to weigh heavily on the debit side."[9]

An even more cautious admonition has been given by meteorologists Chorley and More: "The unknown dangers attendant upon such large-scale tampering with the delicately-balanced world hydrological cycle must postpone such schemes until theoretical mathematical models simulating the behavior of the earth-atmosphere system have been developed, so that all possible effects can be predicted in advance."[10]

Such warnings, however reasonable they sound, must be regarded as unrealistic. Until deliberate climate-changing efforts are made, scientists will never be able to conjure up accurate "mathematical models" of the earth-atmosphere system. The studies of Earth and the accidental manmade meteorological effects can help, as can the studies of the meteorology of other atmosphere-shrouded planets. But the requirement to wait until all possible consequences can be predicted would effectively put off such actions indefinitely—and nobody has ever heeded such warnings before.

For terraforming, the issue of hurricane control is a highly relevant one. For the first time, deliberate climate modification on a large scale is being contemplated without the guarantee of complete success or even complete safety. Lives and property are at stake. Ethical as well as practical rationales are being appealed to—and it is a foretaste of the arguments about rebuilding Mars.

> I was thinking the day most splendid till I saw what the not-day exhibited, I was thinking this globe enough till there sprang out so noiseless around me myriads of other globes.
>
> Night on the Prairies
> Walt Whitman (1819–1892)

5

Resources for Terraforming

The navigator was satisfied. One more hunk of future ocean, she thought, as the symbol on her tracking screen continued to approach the edge of her control zone. As the last human being to pay it any attention for a decade, the navigator checked her trajectory plots for the fourth time.

Wish I'd spent more time on the billiard table in college, she mused. It might have helped! The object whose course she had designed was now headed out on an interplanetary bank shot that would have blown the mind of any billiard player on twentieth century Earth: a slow angle towards Jupiter, then a sudden reverse during a dash deep within Jupiter's gravity, then a renegade retrograde path backwards against the course of all sane planets and asteroids, avoiding Mars and Earth, but barreling straight on in to its ultimate target, 3,866 days in the future, the changeling planet Venus.

Nobody would be hurt in the crash, since there was not a living soul on Venus—*not yet*—*and the payload was guided by a computer almost smart enough to have second thoughts about allowing itself to be destroyed on impact.* The payload was a two-mile-wide tank of metallic hydrogen, mined and refined from the upper atmosphere of the world which lay just outside of the navigator's cabin. *Grandfather Saturn will never miss what we steal if we stay for a thousand years,* rationalized the navigator. *And of course we won't*—*I've got only three left myself, and the project has thirty-eight to run.*

For a moment, the navigator felt an urge to stay a little longer to see the skyhook completed, but she strangled the thought even as it struggled towards birth. *Nothing is going to keep me here beyond my hitch* she promised herself for the ten-thousandth time, not noticing her gradually declining vehemence. There was something seductive about living with a sky half full of a perpetual dawn and an eternal rainbow, something that broke down the resistance of men and women with stronger wills than hers. Half of the engineers on the project eventually extended their ten-year hitches; several had even worked long enough to draw retirement, and one had set up a hermitage in the forward Titanian Trojans, scavenging from the leftovers of the Soviet expedition of '09. *Takes all kinds,* the navigator decided, *but not me.*

The company which had contracted with the Venus Project to deliver hydrogen to the planet, thus absorbing the vast amounts of excess oxygen and providing the planet with oceans needed for half a dozen ecological reasons, had organized its Saturn scavenging expeditions along the lines of a paramilitary frontier camp of nineteenth century Earth. On the first voyage, the flagship *Isaac Asimov* had carried 200 people, mostly married couples, who had all signed on for ten years in space. It had taken longer than that to send the first loads of stolen Saturnian hydrogen headed down into the inner Solar System, and by the time the first loads arrived, a quarter century had passed and few of the original personnel remained. But some of their children now were among the engineers on the much-enlarged (and no longer a spaceship, but a spinning habitat) *Asimov.* And there were grandchildren in the nursery.

The navigator moved to check the payload's trajectory one more time, then held herself up straight. *No, I know it's good. If I check it again I'll never feel confident.* The suppressed realization that the payloads she dispatched along their suicidal falls would, if aimed at Earth, pack destructive punches greater than all of the thermonuclear weapons *ever built or even conceived of—that thought was not one which any man or woman on the staff could grapple with frequently. Time to relax, she told herself. Let's do next month's scooper loading schedules, she decided, turning her attention to the dozen slug-like ships which skimmed across the face of Saturn, trawling scoops into the upper atmosphere to collect the gases needed a billion miles away. The thought of those molecules of hydrogen spending five billion years in sterile suspension, only to be kidnapped and transported to a new world to be eventually ingested by living creatures still warmed the heart of the navigator. I still believe in the project, she realized—yeah, and in the bank account back in Singapore. I believe in that, too.*

On the wall of her cabin was a recent artist's impression of the skyhook which was even now being assembled in the Asteroid Belt. The scoopers would be obsolete, as a single anchor asteroid would be nudged into a close orbit from which lines would be dropped four thousand miles into the Saturnian atmosphere. The hydrogen transfer rate could be quadrupled—fifteen years in the future.

They won't need scooper ship navigators then, she thought to herself. I wonder what training programs I could get into now?

Saturn Orbit, 2199 A.D.

Rebuilding other planets into reasonably Earth-like worlds requires four main resources: first, a planetary body with sufficient mass to hold atmosphere, which will be the object of the terraforming; second, smaller bodies containing volatile materials to contribute to the atmosphere and oceans of the planet; third, steady sources of energy to provide approximately Earth-normal temperatures on the surfaces, along with temporary sources of energy to make the initial changes in the target planet's thermal state and in the physical locations of the volatile materials which are to be imported; fourth, physical and biological agents to effect the gross chemical changes of the atmosphere needed to create Earth-like habitats. In addition, there must be enough *time* for the materials to change temperatures and react with physical, biological, and chemical processes.

The planets themselves will each be addressed in detail in subsequent chapters. Here let us begin by examining the availability and utility of the other resources needed.

Comets

In moments of hand-waving and vivid metaphors, the concept of

104   New Earths

The planets of the Solar System can be rebuilt using raw materials from asteroids and moons, powered by the Sun's energy and by nuclear energy fueled from the atmospheres of the gas giants.

dropping comets on desert planets has been appealing. This process of importing volatiles has been alluded to as "comet-catching."

Comets are actually a poor source of building materials for future Earth-like worlds, however crucial they may have been in the controversial origin of Earth's own volatiles. Comets were larger and more numerous a few billion years ago; today they are probably too small and too unpredictable to be of any practical use.

The total mass of all comets around the Sun has been estimated to

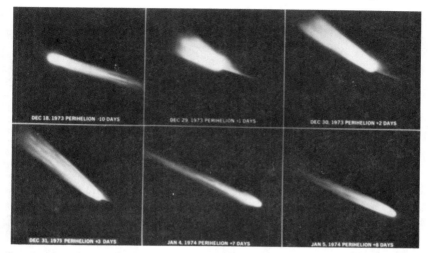

Comets are so spectacular because they are so wasteful with their volatile materials. In fact, they are quite small and hardly worth chasing, except for science.

be not much more than the total mass of Earth.[1] Considering that most of this is effectively forever out of our reach in the distant realms of the "Oort Cloud" (a theoretical source of all comets) there is not much left to work with. Halley's comet, for example, is probably only about 5 kilometers (3 miles) in diameter. Comets larger than 25 kilometers (15 miles) in diameter are expected only a few times per thousand years!

And when they do come in, comets are on erratic trajectories which cannot be predicted in advance. This characteristic makes any attempt to reach a comet and steer it to a desired target planet highly implausible.

Asteroids

Asteroids comprise an important resource for future exploitation, as space prophets have realized for decades. Since the days of Tsiolkovskiy, would-be space industrialists have gazed hungrily at these billion-ton hunks of iron, nickel, silicon, and other industrially useful materials. (President Lyndon Johnson put a governmental stamp of approval on such dreams when, in 1964, he forecast that "the time may come when we will retrieve entire asteroids for mining.")[2] The latest spokesman for these plans is Dr. Brian O'Leary, former astronaut and current ally of Dr. Gerard O'Neill of Princeton.

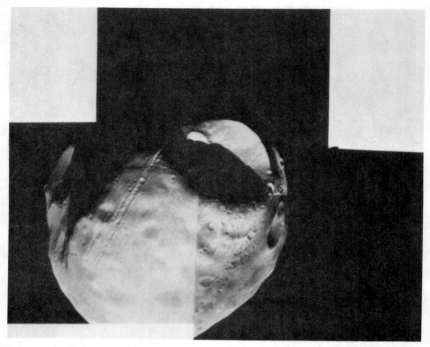

Mars' moon Phobos may be a typical asteroid from the Martian asteroid belt; such objects may provide materials from which atmospheres can be constructed on the Moon and Mercury.

Indeed, the asteroids appear to be ripe for plucking. They come in bite-size nuggets, and they roam through a wide variety of orbits, some quite convenient to Earth. O'Leary has developed detailed plans for the retrieval of nearby asteroids and conversion of their materials into commercially valuable products.[3]

For terraforming, the value of asteroids lies in both their material composition, and their ability to transfer momentum. That is, we may want to drop a few asteroids onto target planets in order to supply them with atmospheric materials. We may also want to drop asteroids onto planets to dig deep craters, or to nudge the planet's spin or orbit into a desired direction.

What are asteroids made of? Prior to about 1970, astronomers who bothered at all about asteroids tried to answer this question from *a priori* grounds. That is, they utilized theories of the creation of the Solar System which were formulated based on the known data about the planets, and asked what such processes would leave over for the asteroids. In addition, astronomers had fallen meteorites in their possession, presumably samples of the asteroids or of some subset of asteroids. Such extraterrestrial samples could provide hints as to how close the earlier guesses were.

Among the dozen separate belts of asteroids, handfuls seem to fall into specific families which seem to point to a common origin in some catastrophic breakup of a larger asteroid. These groupings are called Hirayama Families after the astronomer who first recognized them. They underscore the assumption that the asteroids today are the scattered remnants of a long series of earlier breakup events.

But how large was the original body or bodies? And when did these breakups occur? Although popular scientific literature frequently conjures up the image of a giant tenth planet exploding in some astronomically recent interplanetary cataclysm, the considered opinion of most astronomers is that the asteroids are remnants of planetesimals which may have been several hundred kilometers in diameter, but which were never able to coalesce into a large planet because of the disruptive gravity of Jupiter. In fact, so disruptive was Jupiter's effect that it seems to have nearly prevented the formation of Mars, even further away; Mars might have been ten times as massive, or more, if its intended parts had not been scattered throughout the Solar System by Jupiter.

In the 1970's, *spectral measurements* of the reflected sunlight from asteroids made from Earth have revealed the existence of distinctly different classes of objects, all previously lumped under the heading "asteroid." The measured characteristics were (and are) still unable to provide a definitive explanation of the actual surface composition of these objects, but they did provide criteria for assigning asteroids into these groups or classes.[4]

The dominant class of asteroids are called the *Type C* objects, named for the carbonaceous composition of meteorites which are believed to have come from such objects. About 75% of the main belt asteroids are Type C. These are the objects of most importance for terraforming, since they are the asteroids richest in volatile materials to be used in the creation of an atmosphere: i.e., water, nitrogen, and carbon compounds.

Corresponding to stony-iron meteorites is another class of asteroids, *Type S*. Fifteen percent of main belt asteroids are Type S; they are brighter than Type C asteroids, and probably denser.

Type M asteroids are generally metallic red in color. There are also *Type E* objects with very high albedos and *Type R* objects, of which is known only that they are quite different from all the above. Lastly, there are unclassifiable objects, asteroids which do not fit into this scheme because their light curves simply do not match any other known groups. The Trojans, for example, are such critters; that is, they are all related to each other, but evidently not to any other known asteroids.

Leaving such embarrassments temporarily aside, the classification scheme works very nicely elsewhere. Most important, it reveals a hitherto unrealized fact: the asteroids are not a homogenous mixture of different types, but are instead a skewed population dominated by S asteroids at the inner edge and by C asteroids toward the outside edge. The ratio of S to C asteroids seems to be a direct function of distance from the Sun.

"These distributions verify the increase in relative abundance of the low temperature assemblages with increasing distance from the Sun" writes Dr. Thomas McCord, of the University of Hawaii, a leading asteroid expert.[5] NASA Jet Propulsion Laboratory asteroid specialist Dr. Foster Fanale explicitly declares the doom of the "exploded planet" theorists by proclaiming that "the correlation of asteroid composition with heliocentric distance suggests that, despite considerable scrambling, most surviving asteroids seem (unlike comets) to be in roughly their original orbits."[6]

Presumably, then, the parent bodies of the asteroids formed at the same time that the major planets were forming, but they never seemed to get together. The original total mass of such objects remains unknown, since computer simulations show that over a few billion years, nearly any original total mass (up to a thousand times as great as the present total mass, or even greater) erodes in size from continual collisions until it reaches approximately the present population level—which together does not amount to more than 1% of Earth's mass.

Terraforming may require the importation of materials from the asteroid belt, which is the justification for discussing these subjects. The inner planets do not appear to need additional supplies of stone and metal, although the carbonaceous asteroids may provide valuable water and nitrogen. Thus, the Type C asteroids are of primary interest to us.

Earth-Crossing Asteroids

Moving an asteroid against its will is, of course, a good trick, although a dozen or two different schemes have been imagined. Conveniently, there are already some asteroids which follow paths that cross the orbits of Mars and Earth. Are they potentially usable?

Such objects are called *Apollo asteroids* (no relation to the Moon flights), which cross Earth's orbit, and *Amor objects,* which approach Earth's orbit but do not cross it. They are also sometimes called the *Earth-crossers* and the *Mars-crossers.**

The population of these groups of asteroids can be guessed from those which have already been observed (probably a small but appreciable fraction of the total) and from the cratering rates on Earth, the Moon, and Mars over the last three billion years, caused by the very rare (but catastrophic) collisions of such asteroids onto the planets.

The total number of Apollo asteroids appears to be fairly constant, which suggests that they are being replenished from other regions of the Solar System, while their population is being controlled by collisions with planets. But orbital computations are unable to pinpoint their region of origin: a few may be stragglers from the inner regions of the asteroid belt, but most must be extinct comets—or something else.

The problem for asteroid scientists and terraformers alike is that the Apollo objects do not look like extinct comets are expected to look. In fact, spectroscopic data shows that they are overwhelmingly Type S, which points to the inner asteroid belt as their point of origin, despite the protests of astronomers whose computer simulations cannot explain how so many asteroids stray off course into the inner Solar System. These Type S objects are not likely to be the types of asteroids needed for terraforming; C-type objects have more of the material resources needed, and they seem to be rare among the Apollo objects. So we must go all the way to the main belt.

So asteroids are a tempting source of terraforming material. However, they may not be rich enough or abundant enough or accessible enough. In such a case, we may need to turn to the next nearest sources, the moons of Saturn.

Saturn's Children

Beyond the asteroid belt lie the giant planets Jupiter, Saturn, Uranus, and Neptune with their associated satellites and rings; also somewhere out there are rogue asteroids such as Chiron, discovered in 1977, and the twin worlds, Pluto and Charon, likely to be primarily frozen water and similar volatiles.

---

*Note that just because an asteroid periodically crosses Earth's orbit, this does not guarantee a collision: since the asteroid's orbital plane is usually inclined a few degrees to that of Earth, when the object is actually crossing Earth's orbit, it is either above or below the plane of Earth's orbit, avoiding an intersection.

Saturn and its moons might be the Solar System's most valuable source of raw materials for terraforming the inner planets over the next thousand years.

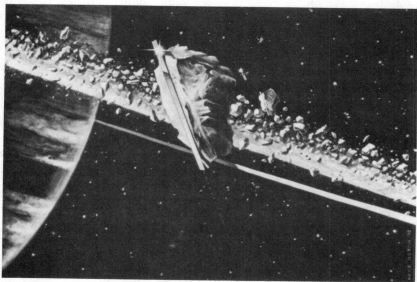

This fanciful view of the theft of ice asteroids from Saturn's rings is well-grounded in science fiction and early terraforming scenarios but is not a likely development. Saturn's rings are too firmly held by Saturn's gravity, while small icy moonlets farther out in space may provide much more likely candidates for exploitation.

Resources for Terraforming   111

We can dismiss the Galilean satellites of Jupiter because of their own significant gravitational forces and the gravity pull of their primary which keeps them locked safely away from our early exploitation; the same arguments apply to Saturn's moon Titan and, mercifully, to Saturn's beautiful ring system, which lies far too deep within the *gravity well* of Saturn to make it an attractive resource for export to planetary engineering sites.

The outer satellites of the gas giants are very tempting. Jupiter's collection of outer moons may be captured asteroids of unknown composition (and to this family add the Jupiter-controlled Trojan asteroids occupying the Lagrangian points ahead of and behind Jupiter in its orbit); Saturn's moons seem to have large portions of water ice; similar compositions probably characterize the moons of Uranus and Neptune (including Triton, but again it's too large for easy export, so it's temporarily safe).

An interesting object to examine more closely is Phoebe. It is Saturn's outermost moon, thirty times as far out as our Moon is from Earth; it takes more than one and a half Earth years for Phoebe to make one "month" around Saturn. Phoebe is about 300 kilometers (about 200 miles) in diameter and may contain large reserves of volatiles.

Other smaller ice moons of Saturn probably exist, and they may be even more attractive as sky water mines. But the situation of Phoebe is sufficient to demonstrate the practicality of such mining, while other moons would make the job even easier.

Phoebe's orbital velocity around Saturn is about 1.6 kilometers per second. To escape from Saturn's grip entirely, Phoebe would need an additional velocity of about 40% of that, or 0.6 kilometers per second. Once free in deep space, it would still be drifting along with Saturn's orbital velocity of 10 kilometers per second; to bring the material in close to the Sun would require that most of that velocity be extinguished, allowing the Sun's gravity to suck the object down into the brightly lit, warm regions of the inner Solar System.

The "Reverse Jupiter Swingby" Trick

There is a clever way to greatly reduce the amount of energy needed to move something from Saturn's orbit into the region of, say, Mars. This would be accomplished by using a "reverse swingby" of Jupiter*, swooping in on a nearly tangent trajectory to within 100,000 kilometers of the giant

---

*This idea, as far as can be discerned, has originated with this author.

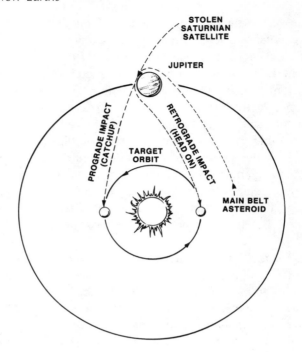

REVERSE JUPITER SWINGBY TECHNIQUE

The "reverse Jupiter Swingby" is a navigation trick for super-efficient transport of material from Saturn to the inner Solar System. The cargo could be twisted onto a "posigrade" or "retrograde" orbit, leading either to an overtaking collison or a head-on collision with the destination planet; both trajectories are useful in different instances.

planet, and having the Jovian gravitational field twist the object's velocity vector onto a path leading straight for impact (with course corrections) on Mars—or any other target. A delicate aim would be required, but we've already done so with the Pioneer and Voyager probes in the 1970's.

This velocity change to reach only to Jupiter's orbit (where Jupiter's gravity would do the rest) is not 10 kilometers per second (the direct-to-Mars requirement), but only 1½ kilometers per second. Added to the amount needed to pull free from Saturn, that makes a total of about 2 kilometers per second. And if the impulse could be delivered at once, the combined Saturn-escape and Jupiter-graze requirements could be accomplished with a single impulse or even less than that, but there may be practical problems in doing it that way. This 2 kilometers per second is the worst case.

However the impulse is delivered, Phoebe would now be on a path towards Mars. Yet what about the small print: how long would it take, and what kinds of restrictions might we be forced to accept due to celestial mechanics requirements—the launch opportunities occur not continuously but very infrequently (the old "launch window" problem)?

The minimum energy path from Saturn-to-Jupiter encounter would take a little more than ten Earth years; from Jupiter, passage to Mars-impact would be about another year. Along the way, course corrections would be needed, controlled either by a robot autopilot or by visiting teams of astronauts. Nobody should be on board for Jupiter fly-by: the Jovian radiation belts would fry any living creatures unless they burrowed a few kilometers into the surface of the cargo-moon.

Additionally, this trick of interplanetary billiards requires that Jupiter and Saturn be in the right initial position so that when Phoebe arrives at Jupiter's orbit, Jupiter is there waiting. Jupiter and Saturn repeat their same relative positions every twenty years, so this planning would have to be made some time in advance.

Using a Jupiter "reverse swingby" maneuver works very well for stolen moons of Saturn, but for more distant planets the bonus is not nearly as pronounced. For example, Neptune's moon Nereid may be considered another mining candidate: its orbit around Neptune ranges from 1.6 million kilometers (1 million miles) out at its close point to 10 million kilometers (6 million miles) at its far point, taking almost exactly one Earth year to complete one revolution. To escape Neptune, an impulse of 0.2 kilometers per second would be needed; another direct fall into Mars, killing off most of Neptune's orbital motion, would require almost 4 kilometers per second more, while the Jupiter "reverse swingby" would still require 1½ kilometers per second (the less favorable reduction is due to the fact that the Neptune-Jupiter transfer orbit is almost as lopsided as a direct Neptune-Mars transfer orbit).

Even so, Nereid is available for less than 2 kilometers per second velocity change, less than the amount for Phoebe. But before we set our sights on Neptune's resources, consider the fine print on this contract in celestial navigation: the Neptune-Jupiter jaunt would take 37 years, almost four times as long as from Saturn. Now, it might be worth it if our future terraformers were far-sighted enough, but, even so, it seems like an unacceptable time penalty.

The tools and techniques to steal entire moons—or just pieces of them—will be detailed in the next chapter. Our concern here is that the *material* we may need to create an atmosphere on the Moon or Mercury, or elsewhere *does exist* and *is* accessible. The wholesale rearrangement of the geography of the Solar System *is* feasible.

Energy and Fuel

Sunlight is the main source of energy in the Solar System, whether in direct form, or through the release of stored sunlight in fossil fuels, or through the mechanical action of wind, currents, and atmospheric electricity. A small fraction of the energy which drives planetary biospheres comes from the decay of radioactive elements, or from the compressional heating of planetary compaction (on Jupiter this form is the dominant contributor), or from the tidal stresses induced by a massive neighboring planet (the best example of this is the volcanic moon Io). But the Sun is the major source of planetary energy.

Solar energy varies by the famous *classical inverse square law:* at twice the distance, only one-fourth as much energy is delivered. Attenuation by dust in the Solar System is not a significant factor, although deliberate human actions to create strategically placed dust clouds near the inner planets may someday be feasible. Conversely, solar energy can be *concentrated* by using mirrors—either for small industrial furnaces or for entire planets.

Recent studies of historical records, and reconsideration of basic assumptions of astrophysics have led to the realization that the *solar constant*—the amount of solar energy received at Earth's average distance from the Sun—may actually vary over a period of decades by as much as several percent. But each year, due to the eccentricity of Earth's orbit, the incoming solar energy already varies by about 5%, so an additional change would need to continue over many years to eventually be felt—unless it were modulated by direct human activities.

Humankind has introduced a new form of energy into the natural balance of the Solar System: deliberately released thermonuclear power, both fission and fusion. Through the last decades of the twentieth century, such energies have remained puny relative to natural forces on Earth's surface. However, within a century or two, it is conceivable that human beings may have the capability to release and direct energy on a scale comparable to a measurable fraction of the Sun's output.

Resources for Terraforming   115

Nuclear Fuels

Where would the fuel for such energy expenditures come from? For direct nuclear fusion such as the activity in the heart of the Sun, hydrogen can be used—and Earth's seawater can provide that. For interplanetary applications, however, other sources would be needed, and engineering considerations might require a different sort of fuel.

During the years-long study project to design an unmanned interstellar spacecraft, the British Interplanetary Society considered different energy plants to power their starship. (Before one suggests that these BIS engineers are wild-eyed dreamers, I must point out that an earlier generation of BIS planners published a design for a manned lunar landing vehicle that was, in its essential respects, identical with the Apollo-11 *Eagle*—thirty years ahead of its time!) The interstellar project (code-named *Project Daedalus*) culminated in a 200-page report released in London in 1978[8] which described the scientific goals of the mission, the instrumentation design, the vehicle design, and the powerplant design, as well as navigation, communications, and control problems.

For propellant, the British team chose helium-3, a light isotope of helium (which normally exists as helium-4). Other energetic reactions using deuterium (hydrogen-2) and tritium (hydrogen-3) were considered and rejected due to engineering constraints on the starship design. (These fuels, however, would be useful elsewhere in the Solar System where great amounts of energy need to be released.)

The Daedalus starship would need thousands of times the amounts of such materials available on Earth. Even the atmospheres of Titan or Triton, if rich enough in the materials, would be depleted by the collosal requirements. So the British team looked to one place in the Solar System besides the Sun where such material existed in sufficient quantity: the atmosphere of Jupiter.

The details of the balloon-borne atmospheric separation plants and of the Jupiter-to-orbit transportation systems have been worked out by the British engineers in fascinating and convincing detail. For terraformers, the lesson is obvious that the gas giant planets (particularly Jupiter) are rich in fuels which may help power unimaginable machines of the coming centuries, when uranium and oil are of concern only to academic historians specializing in a remote and all-but-forgotten epoch of history called "the energy crisis."

The hydrogen in the atmospheres of the Jovian planets is also a potentially valuable resource. On some planets (Venus in particular) our terraforming may create a surplus of oxygen while water is in short supply. This imbalance can be redressed by a wholesale importation of pure hydrogen to react with oxygen, thereby creating oceans.

Genetic Engineering

Where plants and animals are not naturally suited for unearthly conditions on other planets, it is conceivable that more well-adapted forms of these organisms could be deliberately designed in genetics laboratories of the future. Such laboratories might best be located aboard space stations, both to provide a reliable quarantine in the event of accidents, and to allow conditions on the target planet to be easily duplicated for test purposes.

This topic of genetic engineering was also addressed in the NASA terraforming study in 1975.[9] Although the planet in question was Mars, the technique would be applicable to any planet with a surplus of carbon dioxide and a scarcity of oxygen.

Biological Tools

So far, we have moved space mountains and altered the power of the Sun. Adding volatiles may create thick atmospheres and rolling seas, but these worlds will still be sterile and dead, and their atmospheres will be unbreathable by human beings.

Something is needed to convert the material in the atmosphere, primarily carbon dioxide, into free oxygen, using sunlight as a power source. On Earth, of course, biological processes fill this ecological role. Although biology is not noted for its efficiency (it uses only about 1% of the sunlight received by the organism), and although giant chemical processing plants may be envisioned which break down and transform the materials of the atmosphere without resorting to photosynthesis, there is no avoiding the need to create a living biosphere on a planet which people will someday inhabit. An entire ladder of life must be artificially installed and artificially maintained until it achieves self-sufficiency.

The first form of life suggested as a tool of terraforming was blue-green algae (*cyanophytes*), which Carl Sagan once nominated to be the first terrestrial colonists on Venus.[10] According to the NASA study on planetary eco-synthesis published in 1976, such plants would be extremely useful tools in the chemical transformation of a carbon dioxide atmosphere into one of oxygen.

Commenting on the blue-green algae, the report continues: "Their ability to grow or survive in hostile environments and carry out oxygen-evolving photosynthesis suggests that this group of microorganisms should be considered for primary dissemination on a target planet. The wide distribution of cyanophytes on Earth is a reflection of their ecological tolerance and versatile physiology. Cyanophytes are important as primary colonizers on Earth because they are able to grow in environments which are too austere for other organisms, and because their physiological activities bring about conditions which are conducive to the growth of other soil organisms. These activities result in an increase in the humus and combined nitrogen of the soil. . . ." As an example of this adaptability, the report describes how "cyanophytes readily become established on lava which is devoid of other growing organisms" since some blue-green algae species are able to digest atmospheric nitrogen directly.

More research is clearly needed into the genetics of the cyanophyte species, as well as into techniques for inducing genetic changes of any kind. Once that is done, especially tailored cyanophyte strains can be tested in Earth-side laboratories which simulate conditions on target planets.

The goal of genetic engineering is to deliberately change an organism's physical and biochemical characteristics by altering the organism's collection of genes. This can be done by causing small changes in a specific gene or by integrating a set of alien genes into the organism's own genetic apparatus. Evolution has been caused by natural genetic changes; humans also have performed genetic engineering by breeding plants and animals to accentuate desired characteristics.

Recently, however, powerful new biological and chemical techniques have been discovered for altering an organism's genetic design. New genes can be generated from preexisting ones via exposure to certain chemicals, physical agents, or radiations; genes can be physically transferred from a donor to a recipient organism, thus producing new combinations and new types of organisms.

Two techniques for replacing an organism's genes are called *transformation* and *transduction*. The NASA ecosynthesis report defined how they would be used.

> *Transformation* [italics added] is a process in which DNA is extracted from donor cells and taken up by recipient cells. This foreign donor DNA (containing the genes for desirable traits) can be integrated into the genetic apparatus of the recipient cell, thereby endowing these cells and their progeny with the desired trait. *Trans-*

duction [italics added] is a similar process differing from transformation by the use of certain viruses as the donor vehicle by which genes, incorporated into the virus, can be moved into the recipient cells. Recently, a new and another quite dissimilar method has been described. Referred to as *plasmid engineering* [italics added], it utilizes a specific small genetic element, a plasmid, as a vehicle for introducing genes into cells. It differs from the previous techniques mentioned by allowing the use, in principle, of *any* gene from *any* donor. By the use of certain specific enzymes the desired gene is joined to the plasmid and the gene/plasmid complex introduced into recipient cells. Once in the cell the gene/plasmid complex can form a stable, genetically functional unit which replicates and is passed on to the progeny of the original recipient cell.

Stressing that the initial goal of genetic engineering will probably be the development of a special strain of blue-green algae for use on Mars, NASA pointed out that very little research has ever been done on the genetics of such algae, and experimentation is just beginning.

Several techniques associated with genetic engineering have been demonstrated in a few species of blue-green algae, i.e., *chemical mutagenesis* [italics added], *sexual recombination* [italics added], *and transformation with DNA* [italics added]. . . . Another technique, *viral transduction* [italics added], while not yet demonstrated, is quite likely. . . . Chemically extracted and purified DNA, as well as DNA released by cells in growing cultures of blue-green algae, has been used for genetic transformation experiments. Transduction is dependent upon the sensitivity of the cells to a class of viruses called "temperate" viruses. These viruses have the property of acting as carriers of genes between donor and recipient cells. Temperate viruses have been isolated which can infect blue-green algal cells. The transfer of genes between the algal cells by transduction, however, has not yet been reported.

Such information suggests that while many avenues of biological research, which are being pursued for their own special reasons, may be of collateral value to terraforming, it will nonetheless be necessary to carry out investigations uniquely justified by terraforming requirements alone. It may seem a long way from microscopic algae to an entire planet, but they are both part of the same picture.

And After the Algae?

The cyanophyte algae would, of course, only be the first of many strains of terrestrial organisms introduced onto the target planet. Lichens could follow, preparing soil and contributing to the processing of the atmosphere. An entire series of stages must be plotted and modeled on powerful computers, leading to the establishment of a stable biosphere capable of supporting human life.

In Frank Herbert's novel *Dune,* his desert tribesmen on the planet Arrakis are engaged in deliberate programs of reforestation and eco-synthesis. Starting from bare sand dunes, they transform the desert step by step into productive cropland and grazing land.

Herbert, who studied undersea geology, psychology, navigation, and jungle botany before becoming a science fiction writer, learned about desert settlement while doing freelance news feature-writing about seashore ecology in Oregon. This is how he transferred such knowledge into the terraforming of an entire planet:

> Downwind sides of old dunes provided the first plantation areas. The [natives] aimed first for a cycle of poverty grass with peatlike hair cilia to intertwine, mat and fix the dunes by depriving the wind of its big weapon: movable grains.
>
> ... The mutated poverty grasses were planted first along the downwind (slipface) of the chosen dunes that stood across the path of the prevailing westerlies. With the downwind face anchored, the windward face grew higher and higher and the grass was moved to keep pace. Giant sifs (long dunes with sinuous crest) of more than 1,500 meters height were produced this way.
>
> When barrier dunes reached sufficient height, the windward faces were planted with tougher sword grasses. Each structure on a base about six times as thick as its height was anchored—"fixed."
>
> Now, they came in with deeper plantings—ephemerals (chenopods, pigweeds, and amarinth to begin), then scotch broom, low lupine, vine eucalyptus . . ., dwarf tamarisk, shore pine—then the true desert growths: candelilla, saguaro, and bis-naga, the barrel cactus. Where it would grow, they introduced camel sage, onion grass, gobi feather grass, wild alfafa, burrow bush, sand verbena, evening primrose, incense bush, smoke tree, creosote bush.
>
> They turned then to the necessary animal life—burrowing creatures to open the soil and aerate it: kit fox, kangaroo mouse, desert hare, sand terrapin . . . and the predators to keep them in check: desert hawk, dwarf owl, eagle and desert owl; and insects to fill the niches these couldn't reach: scorpion, centipede, trapdoor spider, the biting wasp and the wormfly . . . and the desert bat to keep watch on these.
>
> Now came the crucial test: date palms, cotton, melons, coffee, medicinals—more than 200 selected food plant types to test and adapt.

The Role of People

The levels of human participation in planetary engineering will be many and varied. The actual colonists will be only the final wave of participants in a process which has previously begun. Indeed, the first terraforming teams can be considered to have already formed at the NASA Mars study in 1975, at the University of Colorado at Boulder in 1978–1979, and at the terraforming colloquium in Houston in March 1979.

Science fiction has provided numerous scenarios for terraforming, as already described. In some concepts, masses of people are involved, each using very simple tools: the human wave assault on Venus portrayed in Cordwainer Smith's "When the People Fell," and the fanatical desert tribes conjured up in Frank Herbert's *Dune*. Another approach is the Heinleinesque concept of small independent farmers carving out family homesteads with the help of high technology and energy tools (*Farmer in the Sky*). In Smith's and Herbert's depictions, the people are essentially draftees who suffer heavy casualties; in Heinlein's case, they are idealistic pioneers (but on Ganymede at least, they also suffer heavy casualties due to their dependance on vulnerable high technology approaches). Other colonists have been portrayed as convicts (*Birth of Fire*), as social undesirables (*The Martian Timeslip*), as misfits (Roger Zelazny's "The Keys to December"), and even as religious fanatics. Such models have been drawn directly from episodes in Earth history, enlivened by each author's considerable imagination.

The people actually engaged in the planetary engineering efforts (as opposed to colonization) have rarely been pictured. In many ways theirs is the most interesting role—the building of a planet which they and their children will not live to see. For many, space colonies and traditional domed cities will be their life-long homes.

Once planetary atmospheres have been emplaced, and temperatures have moderated, but before sufficient free oxygen for breathing open air has accumulated, it is very likely that many colonists will already begin to spread across the face of the planet. They will need a modest amount of technology to provide breathing masks and air purification, and either heating or cooling, but the major environmental obstacles to the habitability of the planet—the atmospheric mass and the temperature extremes—will probably have already been overcome. Many of these colonists could be "official" settlers, tasked as farmers, forest wardens, and prospectors; many others could be "unofficial" colonists beyond the

Resources for Terraforming   121

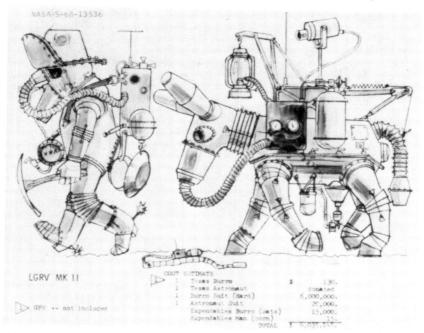

This NASA cartoon contains a kernel of future wisdom. People will use a variety of technology for on-site participants—human and animal—in terraforming planets.

reach of any governmental census roll, living a lean but free existence.

I suspect, in other words, that many people will spill out onto a target planet's surface long before it reaches the Earth-like conditions originally intended for it. And once these people have developed a way of life which compensates for the incomplete atmosphere, they may be less than enthusiastic about seeing the atmospheric engineering projects completed and witnessing the subsequent arrival of hundreds of thousands of new settlers. Such a speculation still remains a topic for an unwritten science fiction novel.

This catalog of resources reads like a "Whole Solar System Catalog"—and indeed, such a book must someday be compiled. One key to understanding the potentials for terraforming is to realize that these resources do not have unique applications to only a handful of projects, nor do they have easily quantified prices. They can be used in numerous different projects, and when the time comes to apply them to terraforming, the developmental costs may have already been borne by other projects which required them first. The whole collection of terraforming resources, then, need not—and likely will not—be built up from scratch, since terrestrial technology will be extremely busy over the next century or two developing capabilities not originally intended for planetary engineering—but extremely valuable for such activities nevertheless. Serendipity—the accidental discovery of value far greater than that originally sought—is a time-honored tradition in science and engineering, and terraforming can hardly be considered an exception.

> *Any sufficiently advanced technology is indistinguishable from Magic.*
>
> "Clarke's Third Law"
> Arthur C. Clarke

6

The Technology

The deacon enjoyed the eternal fireworks on the dark side of the Moon. Each flash symbolized to him another spark of life being accumulated by a world on which he could never walk, but a world for which his children would forever thank him. *Oh, Lord, he prayed,* please let them remember the stars as Thou hast let us perceive them. *But then the ecologist in him dragged him back to practicality.* Oh, well, you've got to have an atmosphere, and it's got to have a greater overhead mass than Earth's, and thus more opacity—we'll send them lots of pictures of stars.

Many of the children were already hard at work at duties in the project, both in **Wilderness** and on board smaller craft closer in to the Moon. The first of them would be able to walk the surface and remove their space helmets within a decade at the rate the volatiles were coming down from Saturn. Already, there was snow lingering on the lunar surface, and gullies forming from flash melting of imported ice.

For the deacon and the rest of the brethren of the Church of the New Covenant, the new world which they were creating would forever remain a promise, but for their children it would be a fulfillment. When they had left northern Europe two decades before, they had vowed to create a new world untainted by the blasphemies and sins of the old. Their generation had set up residence in the space habitat they called **Wilderness**, *there to oversee the metamorphosis of the solid planet and to oversee the education of their children who would settle it. By the time the surface was habitable, twenty years still in the future, few if any of the original emigrants would be alive, so the settlers could begin their new world in a state as close to immaculate conception as was humanly possible. The* **Wilderness** *would be abandoned (would it tarnish their souls if they decided to sell it?, the deacon wondered half in jest), its food-producing "manna machines" brought down to the surface as needed, its crops and herds, which had been carefully bred, transferred to the appropriate climatic zones.*

So many mistakes, the deacon thought with bitter personal regret. So many faulty analogies we tried to make with past experiences. He had been one of the historical experts who had studied the results of earlier frontier societies modeled on religious structures—Deseret, Amity, New Harmony, the Palestinian kibbutzim, *the Bratsk hydroelectric projects, Jonestown, the tragic Olympus-4 space colony, Mars Oases—in order to extract from their experiences a workable guide for the success of the new effort. So many things we should have anticipated—the impact of Earth being always in view; the athletic sub-culture we had a tortuous time stamping out; the clandestine short-wave radio links; the cancerous rise of deviant messiahs.*

Be fair to yourself, his ego urged. And the deacon did grudgingly consider the good news: the efforts which had been denied any chance of success by their mockers back on Earth had finally paid off. For the society from which they had escaped, they had always been throwbacks to the dark ages, a useful target for derisive laughter and cruel physical abuse. Even now, when they were no longer physically present, they were the "loonies," a few tens of thousands of crackpots without whom Earth was way better off. That suits us fine, too, the deacon smiled. Whether or not, as the collegium preached every Sunday, Earth was headed for a planet-wide scouring by celestial fire and brimstone, it was good not to have to share a world with those who were so regimented they hated anyone different. "Conform or perish" had been the ulti-

matum to rebels over most of Earth's long bloody history; only in scattered, blessed times was a third alternative offered: ". . . or get out."

So we "loonies" got out, the deacon recalled. The financing had been the least of their problems, for Earth had grown fat in the harvest of the breakthroughs of the millennium, and energy and food and living-space problems had been finally overcome. The recruitment had been ticklish, though, especially weeding out the numerous converts who began showing up after the existence of the project had become generally known. So many dedicated, talented people to turn down. But one lesson had been clear from the past: any such risky project must have participants already molded into a functioning society prior to actual arrival at the frontier, or jeopardize their chance of survival over the decades. Earth did that for us, the deacon knew. They welded us together before even we ourselves knew the purpose.

It was almost enough for the average Earthman to admit the existence of "Divine Providence," or perhaps even a real honest-to-goodness "God." The deacon, of course, hardly needed such convincing but it was still nice to know.

The Moon, 2288 A.D.

In earlier chapters we have seen a lot of "hand waving" and "wait-until-later" appeals concerning the actual *details* necessary to rebuild the Solar System. It is now time to make a list of the kinds of tools and techniques which are within our comprehension at the close of the twentieth century. Even though it should be undeniable that additional discoveries will be made over the next few centuries, it will be very satisfying to know that the capabilities already within reach are probably sufficient to begin planetary engineering on a large scale.

Several different kinds of activities will be involved. In some cases, material must be transported from the surfaces of large planets; in other cases, very massive objects must be moved about across the entire width of the Solar System. In addition, energy fluxes must be modified, sometimes to increase the level of sunlight on a target planet and sometimes to decrease that level; magnetic fields and radiation belts must also be manipulated.

All of these tools and techniques may be needed in various com-

126 New Earths

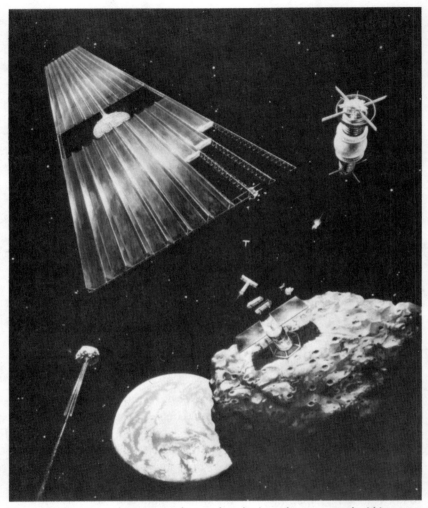

The "mass driver" machine may replace rockets for interplanetary travel within twenty years, and in larger versions can actually move entire asteroids from planet to planet.

binations and sequences, depending on the initial conditions of the target planet, the time scale required for its transformation, and the level of fidelity of the duplicate Earth to be created. So let us examine each element in turn, realizing that this is only the most basic terraforming arsenal, and that new ideas are continually being created.

Terraforming Cargo

The relative merits of different transportation schemes depend, of course, on the type of cargo being transported, which determines the allowable acceleration, the amount of life-support needed, and the maximum tolerable travel time. For terraforming, several widely different types of cargo are involved: people to actually carry out the planetary engineering; more people to act as colonists on the terraformed planets; automata (robots and similar computer-controlled machines) to carry out tedious or dangerous work; machines of all kinds to be the tools of terraforming, ranging from micro-organisms to shovels to H-bombs to giant reflectors; and raw materials, ranging from shiploads of trace minerals to 100-kilometer-wide asteroids and moons.

Each different payload will entail using a different combination of techniques. The human beings may require relatively fast travel times (although the large numbers of colonists may be assigned a transportation system characterized by its cheapness, implying long, slow interplanetary coasts). An ideal propulsion system would be one which can maintain a constant accleration of *one G*, equivalent to the force of gravity of Earth's surface, over a *long* period of time (hours or days, as opposed to minutes as is possible today). With such a capability, a spaceship could reach the Moon in two hours, could reach Mars in two days, and could visit the outer limits of the Solar System within two weeks. The design of such a propulsion system remains a task for the next century. It is mentioned here only as a demonstration of the "speed limit" imposed by the human body's inability to tolerate accelerations higher than one G for long periods of time.

The gentlest possible interplanetary trajectories are those in which the cargo is launched from Earth and then slowly builds up speed using an ion rocket or, more useful for terraforming, a giant solar sail which later could double as a reflector of sunlight. Such interplanetary journeys would be controlled by robot autopilots since they would take months or even years to deliver their cargoes of machinery and non-perishable supplies.

How to Move a Moonlet

Similar travel times are probably in store for the stolen asteroids and moons described in the last chapter. At the smallest possible cost (and considering the mass of the objects involved, even the "smallest" cost is likely to be gigantic and, hence, not willingly exceeded), the moon Nereid would take forty years to reach Mars or Mercury. Saturnian satellites would need two decades even after they have been nudged free of their parent planet. Asteroids could be transported from their native belt to an impact with their new home planet within a period of several years.

Using the pressure of pure sunlight, a giant "solar sail" tacks its way close to a target comet. After attaching lines to the object, the robot-controlled sail could tug the comet anywhere in the Solar System.

The purpose, recall, is to use the volatile materials of the objects to build up the atmospheres and oceans of the target planets. This implies that the object to be moved will be fairly fragile—and this adds further restrictions to the available techniques for moving them around the Solar System.

Three techniques appear to be potentially applicable: light (or *solar*) sails, which could tow the objects on long lines; *mass drivers* (essentially electromagnetic catapults which throw hunks of material backwards using a mechanical means, thus producing a forward thrust on the whole structure) which could use part of the body itself as reaction mass; nuclear blasts, which could knock the body onto a desired course violently by propelling portions of it in the other direction.

Light sails, needing sunlight to obtain thrust, rapidly loose their effectiveness in the outer Solar System where sunlight is weak (unless giant lasers are used to illuminate them).[1] Yet this is where the most important resources are located. Potentially useful as far out as the asteroid belt, where sunlight is one-tenth as strong as near Earth, light sails do not appear to have much future beyond Jupiter.

Mass drivers have already been described in great detail by several space experts and by Dr. Brian O'Leary, an ex-astronaut who is now developing plans for mining nearby asteroids.[2] Such proposals have considerable attractions purely on their own merits, but are even more significant because pursuit of such programs would develop the implements which could subsequently be applied to planetary engineering.

O'Leary's plans suggest that up to two-thirds of a target asteroid would have to be "thrown over the side" to obtain the action-reaction effect needed to propel the remainder into a region near Earth where it could be mined. With more efficient mass drivers, such a fraction of loss could be reduced.

There is still an upper limit to this system, and involves the source of the energy used in expelling the reaction mass from the ends of the miles-long mass driver tubes. In O'Leary's compelling scenarios, the electrical power comes from sunlight collected by giant solar panels.

For applications beyond the asteroid belt, such a system once again runs head-on into the inverse square law which determines the weakening of solar energy as a function of distance. By the time we move such a system as far out as Saturn, it would have to survive and operate on sunlight 100 times weaker than that in the proximity of Earth.

Other sources of power are, of course, possible. Nuclear fusion reactors can be predicted. Indeed, the weightlessness and vacuum of space may form an extremely hospitable environment for nuclear fusion power plants—and many people on Earth may be entirely satisfied in banishing such machines far beyond the limits of the atmosphere. Such an eventuality could have extremely beneficial implications for providing power for planetary engineering.

But in the absence of any serious expectation of such developments, our need to move asteroids and moons is left with one alternative: thermonuclear blasts. Although the notion of our great-grandchildren zipping around the Solar System tossing H-bombs out their spacecraft windows is not especially appealing, many of the superficial drawbacks of such a tactic are not as serious as may appear at first glance.

As early as 1968, people were studying the use of nuclear blasts to divert asteroids.[3] That was the year that the asteroid Icarus made a moderately close approach to Earth, and as an exercise in astronautical engineering, a class of students at the Massachusetts Institute of Technology prepared a plan for diverting the asteroid in the hypothetical event that it would have turned out to be on a collision course with Earth. This project led to the publication of their final report, *The Icarus Project*, by MIT Press. More than a decade later, the concept came to simultaneous fruition in several Hollywood movies on the same topic. (In fact, one of the repeatedly rejected and ultimately abandoned scripts along those lines belonged to this writer!)

Radioactive contamination of the target body is not as bad as it might seem. The use of thermonuclear blasts would not be likely to produce either radioactive cargo or contamination of any region of the Solar System. First, the nuclear devices (and this is more than just a euphonism for *bomb*, since the devices will be optimized for their purposes and would probably make very poor weapons) will be clean with respect to the production of radioactive fission products from the device itself. Secondly, since the target object will consist primarily of volatile materials (afterall, that's why it was picked), there will not be large concentrations of heavier elements which could be converted by the nuclear reactions of the blast into radioactive isotopes. Additionally, the volatile material will boil off quite easily, shedding any radioactive material which still might contaminate it.

Such blasts, then, could indeed deflect the target body. This is not to say that it will be done this way, but it is thought-provoking to realize that large objects can be moved safely and reliably to desired locations in space with technology existing today.

And When They Arrive...?

Years after beginning their journeys, the stolen moons and asteroids approach their destination. Perhaps along the way, they were refined by giant robot-controlled machines, their volatiles neatly frozen, and their solid materials sorted and discarded. Perhaps, in order to preserve their volatile components from the ever-increasing heat of the Sun, they were enveloped in a molecule-thick layer of reflective aluminum or other metal, possibly even mined from their own material.

The easiest way to reach the target planet is to aim straight in and smash to the ground; the asteroid's terrible momentum would be converted to heat which would vaporize its material and would dig a deep crater on the target world. If such craters are tolerable, and such flash evaporation is desirable, the transfer trajectories can be especially shaped to maximize impact velocities.

The most obvious way to do this is to use the gravity of Jupiter to reverse the asteroid's orbital course entirely, twisting its originally prograde trajectory onto a retrograde one whose orbital motion is opposite in direction to the motions of the planets. The last chapter showed that a Jupiter swingby trajectory was a very productive technique. A slightly closer pass to the gas giant would be enough to completely turn the object's orbital motion and put it on retrograde course. Such a fate has already befallen innumerable comets over the billion year history of the Solar System. Humankind would only be duplicating what nature has already done.

The retrograde orbit would mean that impact velocities on the target inner planets would be several times as great as they would be for objects on prograde paths. The collisions would be head-on rather than catching up. Since momentum is a function of the square of the velocity, the energies released from the same initial mass would be ten times as great or even more. This energy would go into heating the planet, a desirable trait on Mars, or into excavating the ground and mixing it with water, a potentially useful technique on Mercury and the Moon.

Sometimes it may be necessary to bring the incoming asteroid down *gently*. Perhaps colonists are already on the surface, or perhaps a delicate biosphere has just begun to function.

In such a situation, an overtaking trajectory is clearly the best. But instead of aiming for a direct collision, the asteroid is aimed about a million kilometers away from the target planet. At this closest approach, a series of thermonuclear blasts slows the forward velocity of the asteroid, allowing it to be captured by the planet's gravity.

Then the asteroid could be slowly worked in closer to the target planet, perhaps by additional nuclear blasts, perhaps by the acceleration of giant solar sails which would work quite nicely here in the inner Solar System. Eventually, the asteroid would be only a few hundred kilometers above the surface. At this point, the reflective protective coat would be removed and the Sun's heat would begin to evaporate the object; the solar wind would drive the material away from the Sun, right into the upper atmosphere of the target planet. On half of the orbit, when the asteroid was "down Sun" from the planet, it would also be in darkness, so its material would not be lost into interplanetary space. At the same time, mass drivers on the asteroid could be shooting bite-size chunks of the asteroid directly into the atmosphere where they would vaporize completely.

In the end, either violently or gently, the volatile material of the asteroid would contribute to the new atmosphere of a planet undergoing terraforming. The asteroid's gases, frozen for billions of years, would soon fill the lungs of human beings.

Skyhooks

"Hitch your wagon to a star" used to signify unbounded and unfounded optimism. In the next century, it may achieve literal fulfillment as "space elevators" carry cargo from Earth up to space stations—and beyond.

Skyhook was once an imaginary machine of legendary and practical jokes, from the same armory which produced "left-handed blivets" and "smoke shifters" and similar non-existent devices which old-timers would delight in sending naive rookies out in search of. But human imagination will soon overtake the fantasies, and someday a clever young engineer will actually build a skyhook, much to the consternation of his mocking seniors.

To space engineers, a skyhook now signifies a cargo line dropped all the way to Earth's surface from an anchor satellite hovering 40,000 kilometers above the equator. Trolleys powered by space-generated electricity can run up and down the line, carrying people and goods into space at extremely low cost.

Such a low cost journey would not be free by any means. Lifting the cargo would lower the space anchor proportionately, requiring the use of some propulsion scheme to raise it again. That could be done; the major practical roadblock to skyhooks is that the line material must be hundreds of times stronger than the best construction material now available.

A clever variant of the anchored skyhook was developed by Hans Moravec in 1976–1977.[4] He called it the "nonsynchronous skyhook," and pointed out that material strengths *already feasible* could support such a device on *all* the planets and moons smaller than Earth.

The Moravec Skyhook consists of a shorter (thousands of kilometers versus tens of thousands) cargo line which is revolving around a rotating anchor satellite, much like the spoke of a turning wheel revolves around the hub. The anchor satellite is placed in an orbit over the target planet, at an altitude equal to the length of its cargo line. It then begins spinning in the plane of the orbit with a period the same as its planet-circling period.

The result is that the end of the cargo line touches down on the surface of the planet at the same point, once each revolution. Cargo modules could be detached and new ones attached. The local motion of the skyhook is almost entirely vertical: it descends from the sky, pauses, and then rapidly retracts back into the sky.

Once again the energy is not free, and must be compensated for by the application of thrust to the anchor satellite. But the system is so efficient that low thrust propulsion is adequate.

Variations to the basic system are possible. If the anchor and cargo line were set revolving at *twice* the standard rate, the cargo hook would touch the surface at two points on opposite sides of the planet every single orbit. More frequent multiple touchdowns are entirely feasible.

If the cargo is released from the line as it is on the upward leg of the rotation arc, it would be flung off into deep space with considerable force. The anchor station must compensate for this energy but would have a propulsion system chosen for its efficiency in meeting these needs.

Skyhooks such as these are, therefore, of great potential utility in transporting mass up out of a planet's gravity field. That mass could be machinery, raw material, fuels, or people bound for deep space or for planets being rebuilt.

Aerostats

Lighter-than-air vehicles have been around for a long time, but in 1979 several architects and engineers proposed an essentially novel kind

134 New Earths

Aerostats, giant free-floating structures, may be built on Earth in the next twenty years. Their most important applications, however, may be as base camps in the atmospheres of Venus, Jupiter, and Saturn. *Art courtesy of Aerostat.*

of buoyant structure that has an exciting potential for use both on Earth and on other thick-atmosphere planets, such as Venus, Jupiter, and Saturn.

Their system was called *STARS*—"Solar Thermal Aerostat Research Station." In a scientific paper released by authors E.C. Okress and R.K. Soberman of the Franklin Institute of Philadelphia, the whole vehicle is described as an up-to-one-mile wide, constant-volume, rigid hot air balloon designed to hover 20 miles above any spot on Earth. Solar energy is to be collected and used to heat the entrapped air—a difference of just 150°F between outside and inside air temperatures—which would allow the giant aerostat to support a payload fully as heavy as its own basic structural weight. The technological key to such a capability is in designing the hull out of lightweight structure and material; the most promising design uses Buckminster Fuller's *tensegrity* (for tension-integrity) geodesic sphere overlain by Tedlar film .001 inch thick.

This STARS aerostat is a machine, extracting daytime energy from sunlight and storing some for use at night. Without proper functioning of its engines it would gradually lose altitude. But the efficiency of its propulsion system is so great that the vehicle could float at the edge of space for years, if not indefinitely.

On Earth, such giant aerostats could be used for weather modification, for spacecraft launch and recovery, for scientific observations, and for many other purposes. These same purposes could be served on other planets, too.

There is one particular problem with aerostats: how does one assemble such a vehicle and get it into position? On Earth, one good way seems to be to build it underwater, with just the sections being worked on jutting out, until the water is driven out by hot air and the aerostat rises like a bulbous Aphrodite from the sea-foam. Land assembly and mid-air assembly have also been investigated by the authors of the STARS system. But none of these techniques would work on other planets whose surfaces are inaccessable.

In the January 1980 issue of *The Construction Specifier,* one other possibility is mentioned: orbital assembly.[5] The authors mention one insurmountable snag: "Unfortunately, it is impossible to lower a space-assembled aerostat into the atmosphere without destroying it." That simply won't do. With whole other worlds to terraform, from footholds aboard aerostats in the upper atmospheres of these potential new Earths, we'll have to think harder and devise a technique which will allow space-assembled aerostats to be lowered safely into the atmosphere of Venus or Jupiter. Well, we have about a century to figure it out.

A string of mirrors in orbit, as envisaged by NASA's SOLARES project, could beam sunlight down to collection points and change the planet's whole climate.

Doing It With Mirrors

One of the potentially most powerful machines for the alteration of planetary energy flow is a simple mirror in space. Mirrors could concentrate sunlight on specific regions of the planet, or could act as parasols and deflect incoming sunlight, in effect, giving the planet a higher albedo.

Space mirrors seem first to have been mentioned by the German

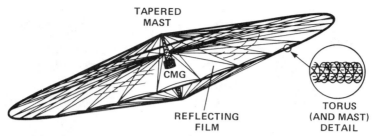

This extremely promising space tool is actually quite simple, yet it may allow the rebuilding of other planets within a century.

rocket prophet Hermann Oberth in the 1920's. Inspired by the legend of Archimedes who used concentrated sunlight reflected from the shields of his city's warriors to set fire to the sails of beseiging Roman galleys, Oberth suggested that orbiting mirrors would be effective war weapons, burning down enemy cities, forests, and croplands at will.

More peaceful applications have been envisaged by another German space visionary, former V-2 engineer Dr. Krafft Ehricke. Testifying before a special Future Space Programs Congressional panel in 1975, Ehricke described a series of stages in his *Space Light* project (or the more poetically named "Lunetta" project).[6]

The "Lunetta" project would create a string of mirrors with a total area of about 20 square kilometers (8 square miles), placed in a special orbit always in sunlight. The light from these mirrors would be up to 700 times as bright as the full Moon, and could be used for night time illumination for search-and-rescue, agricultural, and routine security operations.

A larger system, consisting of either more mirrors or bigger mirrors or a combination, would be part of the "agrisoletta" system. Ehricke suggested using such lighting to increase kelp growth in the ocean in order to harvest methane as an industrial fuel. A larger system in the "agrisoletta" project, with a total of 3,000 or 4,000 square kilometers (about 1500 square miles) of reflectors, could be used for comprehensive, local weather stabilization, such as preventing sudden frosts over citrus groves.

An even larger system called the "powersoletta" would have more than 10,000 square kilometers (4,000 square miles) of mirrors, enough

138 New Earths

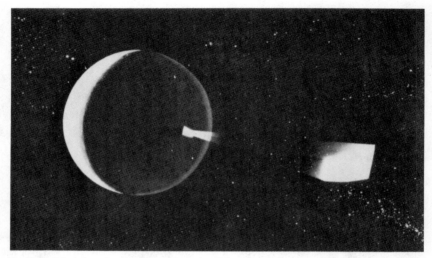

Krafft Ehricke's imaginative but entirely feasible "biosoletta" concept could strategically alter climate over whole states or regions for the benefit of agriculture. *Art courtesy of Krafft Ehricke.*

to provide daytime-equivalent illumination to locations on the dark side of Earth. The primary use of such mirrors would be to allow 24-hour functioning of solar power systems on Earth's surface. NASA engineers at the NASA Ames Research Center refined this system under the code name *Solares* in 1977–1979.

Each of Ehricke's "powersoletta" mirrors would have measured about 5 kilometers (3 miles) in diameter. Only 1 in 8 of them would be in range of the ground station at any one time, but far from being a waste, this factor would call for the establishment of a number of similar ground stations in other parts of the world to utilize these out-of-range mirrors. Thus the very engineering nature of the "powersoletta" system would require that it be international in conception, implementation, and operation.

The energy delivered to Earth's surface by such space mirrors could, within a few decades, become an appreciable fraction of the natural input of solar heat. Localized, such energy could influence evaporation, ice and snow melting, and wind formation. Regionally, such mirrors could steer hurricanes or heat the polar regions to a point where such natural thermal redistribution systems are no longer needed.

Their use in future terraforming efforts on Earth makes them worth studying, if only to avoid all-too-typical climatic miscalculations. But space mirrors have a great future elsewhere in the Solar System since they won't even need transportation systems: they can fly themselves, using the pressure of the solar wind. The smallest Lunetta versions are about one-fifth of a square kilometer (100 would be needed); Ehricke sees mirrors up to 50 square kilometers for this "Biosolletta" application (2,000 would be needed). Larger ones are possible, but control problems and bending may limit the ultimate size of space mirrors, requiring that additional capacity be obtained through buying additional mirrors, not building one bigger.

As Ehricke sees it, these mirrors will be based on rigid structures of sodium-covered graphite epoxy and carbon fiber; the reflective membrane will be sodium-coated mylar or kapton material. These are, materially at least, good choices; but the elements to be used are rare in near-Earth space resources such as lunar soils, so they would have to be hauled from Earth or mined in the asteroid belt or beyond. The discovery of non-terrestrial sources for large amounts of sodium, carbon, and other space mirror constituents is therefore an urgent issue for would-be space industrialists and planetary engineers.

Maintenance of these space mirrors, as Ehricke foresees, will require that they be recoated with sodium (a space process called *vapor deposition,* in which the atoms of sodium are splashed directly onto the surface, could do the trick) at ten year intervals. The mirrors, strung out in chains along specific orbits a few thousand kilometers above Earth could, by the use of the sunlight pressure falling on them, "sail" practically anywhere within the inner Solar System at practically no cost in fuel.

Reducing Sunlight

The use of giant space mirrors to concentrate weak sunlight onto a planet farther out than Earth makes good sense, as already described. The opposite problem is encountered with planets closer in to the Sun than Earth: there is too much sunlight, twice as much directed at Venus and ten times as much directed at Mercury. If those worlds are ever to be terraformed, techniques will be needed to sharply reduce the incoming solar radiation.

Perhaps the same mirrors used over Mars would be sufficient over the inner planets, but they could be aimed differently, casting shadows rather than harvesting surplus sunbeams. Orbital mechanics, however, require that the mirror circle the planet periodically, and it will be be-

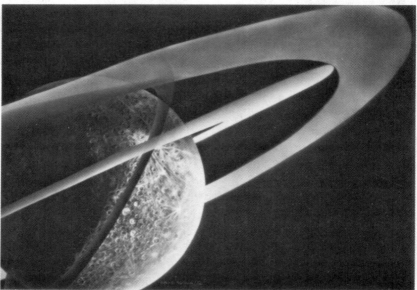

Two ways to shade a near-Sun planet are shown here: first, an artificial dust cloud can be created at a point between the planet and the Sun; second, rings of stony particles can be created in close orbits at various inclinations, casting a shadow on up to 25% of the planet's surface.

tween the planet and the Sun for only a small fraction of each orbit.

This suggests the placing of many objects into an orbit around the target planet, each taking a turn in shading the day side. Such a technique was proposed by Chorley and Moore in the mid 1960's when they wrote that regions on Earth could be shaded by particles placed in synchronous orbits above them.[7] Unfortunately, this suggestion revealed a complete misunderstanding of such orbits: they are possible only over the equator, and, because of the tilt of Earth's axis, the shadow cast by such an object passes either over or under the Earth when the object lies between Earth and the Sun; perhaps twice a year, the shadow would sweep across the day side of Earth for about one hour. Clearly this is not an efficient way to shade a planet.

If a solid orbiting object will not stand in one place long enough, perhaps the problem can be overcome by spreading opaque material entirely around an orbit, encircling the planet and creating a ring like those of Jupiter, Saturn, and Uranus. Jupiter's ring is in the same plane as the planet's orbit, so it catches sunlight edge on and creates no shadow worth mentioning; Uranus' ring is tilted nearly at right angles to the planet's orbit, so its shadow generally does not even fall upon the planet; Saturn's ring is tilted about 15° to the planet's orbit, and it does indeed cast a shadow, darkening a small but appreciable fraction of the day side of Saturn.

Ring Around Earth

As further testimony to the planet-wide climatic effects of a ring system in orbit, it was suggested in mid-1980 that Earth itself may once have had a temporary debris ring which caused a hemispheric ice age and the extinction of numerous species.

Dr. John O'Keefe of NASA's Goddard Space Flight Center postulated that this mysterious ice age 34 million years ago—during which temperatures in the northern hemisphere dropped by 35°F—was caused by a ring of debris which arrived from space, possibly from a volcanic eruption on the Moon.[8] A major tektite meteorite fall occurred at that time, and O'Keefe proposed that other material from the same extraterrestrial source went into making the ring. This ring could have lasted for tens of thousands of years before gradually dissipating.

Whether or not O'Keefe's postulated tektite ring was the cause of the Eocene climate disaster (and the suggestion strikes most geologists and paleontologists as imaginative but far-fetched), the fact remains that rings around planets—whether natural or artificial—can have a profound cooling effect on that planet's climate.

Saturn does not need shading, but Mercury and Venus do. Rings like those of Saturn, perhaps at several different inclinations, could cast shadows on a fifth or a quarter of the day side of the planet, reducing total sunlight accordingly.

The creation of such rings would require the placing of a small asteroid into orbit around the planet, and then pulverizing it. The combination of techniques already mentioned which would be needed to carry out such a planetary engineering task is an exercise left to the reader.

Between the planet to be shaded and the Sun lie several interesting points of space, defined by orbital dynamics. One is called the *first Lagrangian Point*. The fourth and fifth Lagrangian points, also called *Trojan points*, lie ahead of and behind the target planet, each forming an equilateral triangle with the planet and the Sun. Another is the neutral gravity point at which the Sun's gravity balances that of the planet.

Already, artificial satellites have begun to make use of these locations. In 1978, a small Explorer satellite was launched by the United States into the region of the Sun-Earth Lagrange Point #1 (or "L-1"). The probe could have been placed exactly at the point, but that would have located it right in the middle of the Sun as viewed from Earth which would have caused its small radio beacon to be drowned out by the static of solar activity. Instead, the Explorer satellite was guided to slowly circle the L-1 point, watching for oncoming shock waves from solar storms and relaying warnings back to Earth.[9]

Other artificial objects could use that point in space, and they could be placed exactly on target so as to deliberately cast shadows back onto the target planet. Large parasols or artificial dust clouds would both serve. Navigation problems would be serious, and the solar wind would tend to sweep the dust clear in a matter of months, so continued maintenance of the equipment would be required. But it could be done.

And since the inclination of the orbits of the inner planets to Earth's orbit is a few degrees, the shadows cast by such constructions would not affect Earth's sunlight. Even in the worst case, during solar transits by the planets in question (which occur several times per century), the sunlight received on Earth would fall off only slightly for several hours.

There are, then, numerous ways to reduce the amount of sunlight falling onto planets. Such techniques will be needed for the rebuilding of Venus and Mercury, and someday such techniques may even save Earth's biosphere from solar fluctuations which threaten to initiate a "runaway greenhouse effect."

Moving Planets

We've saved the best for last. All the previous tools and techniques are very likely to be used in terraforming in the next few centuries. Rearranging planets is not as popular in terraforming schemes, but is included here to demonstrate that even the most incredible feats may one day be possible.

The concept of altering the orbits of comets was a common theme in the late nineteenth century, and by the 1930's Olaf Stapledon was talking about (but not describing) the process of human action to move planets closer or farther from the Sun. Tsiolkovskiy had speculated on the same subject at the turn of the century.

The main motivation behind such an effort, apparently, would be to alter the amount of sunlight reaching the planet. Yet as we have already seen, there are numerous ways to do that artificially without the energy expenditures required to move whole planets: sunlight can be increased by the utilization of large mirrors, without practical upper limit; sunlight can be lessened by the use of parasols, dust clouds, or similar blocking material.

But suppose there was some specific requirement which could not be solved without the physical relocation of the planet in the Solar System. Could it be done? How might it be done?

Scaling up the traditional techniques mentioned earlier, such as nuclear blasts, solar sails, and mass drivers, we soon reach some fairly solid limits. For example, a planet with an atmosphere would effectively frustrate such techniques; planets a thousand miles or more on a side cannot be hauled around by solar sails much smaller than the radius of the Solar System itself, in any "reasonable" length of time.

Strictly as a demonstration proof, let us examine two possible techniques for moving large Ganymede-size planets, with the complications of thick atmospheres thrown in. These two techniques, which I call "multi-swingby" and "hosing," are not to be seen as a forecast of what will someday take form, since I expect that far better ideas will be devised in the next few decades. Rather consider them as demonstrations of the fact that, even within the limits of today's imagination, moving planets around is *not* an "absolutely impossible" task.

The "multi-swingby" technique takes advantage of the fact that two bodies passing close to each other exchange momentum, a dynamical process first utilized in the Apollo program in 1968 to fling spent S-4B

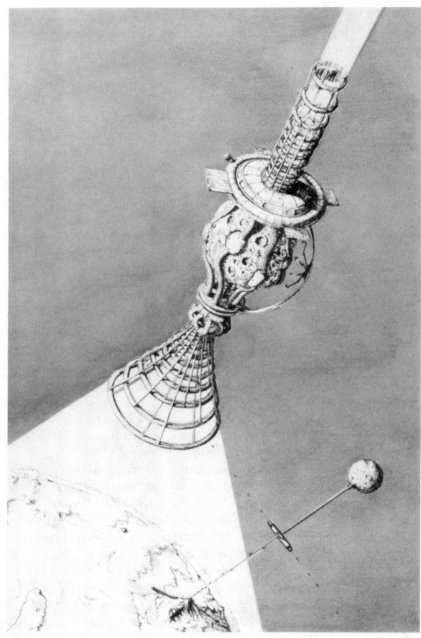

This highly fanciful (but entirely conceivable) machine could be 50 kilometers (30 miles) long, and could move planets into new orbits over a period of decades.

boosters past the Moon into solar orbit. Later it was the standard technique for the Pioneer Jupiter probes and the Voyager "grand tours." Reverse Jupiter swingby maneuvers have been described for transporting Saturnian volatiles to the inner Solar System.

Theoretically, a certain desired velocity change could be imparted to a target planet via a single swingby of a large enough dummy mass—which I choose to refer to as the *cue ball*. But there are many practical problems with this approach. First, to move to a new circular orbit from an old one would require *two* separate impulses, so the maneuver cannot be done with a single pass, no matter how powerful. Secondly, the gravitational stresses of one large body on another are bound to raise tremendous tides on the target planet, which would be hazardous to passengers. Lastly, the task of maneuvering the cue ball, which is going to be of at least equal mass to the target planet, would simply raise again the basic, original problem of moving a large planet.

For all these reasons, a succession of small cue ball near-fly-by maneuvers seems most appropriate. A manageable object, say a few hundred kilometers across, could whip past the target planet at high speed (preferably retrograde, so as to allow repeated encounters several times a year for each mini-cue ball) and impart a tiny gravitational nudge—a nudge so gentle as not to be detected from the target planet's surface except by sensitive instruments.

We have no precise numbers to offer on this scheme. Rough calculations suggest that moving Earth's orbit out from the Sun by 10% would require several hundred thousand such maneuvers. Since each cue ball could, at most, handle two encounters per year, to carry out this scheme in less than a century would require more than 1,000 cue balls, resulting in encounters every two or three hours or so for 100 years. The mass of cue balls to be juggled would amount to enough for a moon-sized planet all their own—not a resource likely to be available unless the project priority were very very high.

The "hosing" method involves building a series of giant propulsion machines which would spray a diffuse stream of water onto the backside of the target planet, imparting some momentum (and a lot of random air motion). An equal propulsive force must be applied in the opposite direction to keep the planet pusher in position behind the target planet. And since this gentle but inexorable push must be maintained for decades as well, a scheme must be developed to remove the excess water which has collected on the planet from the hosing. Such a method could possibly involve productive application of the skyhook system mentioned earlier, to haul extra water up off the planet and transport it back to the planet pushing machine(s) also.

The system in action would then look like this: a typical pusher machine would station itself perhaps 30,000 kilometers (20,000 miles) behind the target planet (say, terraformed Venus), its forward projector splayed out like a medieval blunderbuss to spray an entire hemisphere with dilute reaction mass and its aft projector straight to project as narrow a beam as possible, so as not to interfere with interplanetary commerce. Extra reaction mass—perhaps an imported asteroid—is attached to the machine (or perhaps the modular machine is built around an asteroid which is to serve as reaction mass).

The method of propelling the reaction mass could be any of the schemes listed earlier, or something entirely novel. Whatever it is, it will obey the action-reaction requirements formulated by Newton more than 1,000 years before. One stream of material spurts out to infinity, while the other impacts the target planet, imparting to it a gentle push.

The refuelling system which recovers the water sprayed onto the planet is left to the reader's imagination. One thing must be stressed, however, in this apparently closed loop system: it is not free. Every step of the way requires energy to propel the material and to then extract it from the planet's gravitational field for reuse or for random disposal. That energy goes into "hosing" the planet farther into space—or into slowing it down so it can be dragged close to the Sun.

Theoretically planets *can* be pushed around. The technology can be imagined, while the energy and time requirements are stupendous. For the first few centuries at least, maybe our mirrors and parasols offer the most attainable goals.

> Brothers. . . . Stoop not to renounce the quest of what may in the Sun's path be essayed, the world that never mankind hath possessed.
>
> *Ulysses*
> in Dante's *Inferno*, XXVI

7

Mars—A First Look

The mountains here were higher than in the land of his birth, but the village headman regretted that they were not as rugged. He vaguely missed the skyline of the Andes, and in moments of irrational nostalgia, he was sorry that his bones would rest on a world his ancestors had once worshipped from afar but had been too wise to ever dream of visiting.

What were the reasons? he wondered to himself again. His wife had known all of them, and while she lived he, too, could recite them. But now she was buried on a small hillside in a cemetary that looked strange because it was so new and still so sparsely populated. It didn't used to be like that back on Earth.

That was a reason which even his wife in her wisdom had not been able to formulate. Twice his native village had been the scene of wars; he had lost count of the bandit raids and the subsequent famines. That is a difference. He groped for the thought—we do not bury our children any longer.

Of course the young people had mouthed other reasons during the long debate when his brother had returned from the lowlands with the news that whole villages would be recruited in their entirety for emigration to the newly opened highlands of Mars. For the young, it was excitement; but mainly it was a way out of the trap which they saw for themselves. They did not want to grow up to be like him, the headman knew, and he knew that they had been correct. Their hopes—the first hopes that his village had believed in for fifteen generations—had been irrational, but their gamble had paid off.

Or was it a gamble? They had certainly worked for it. First was the screening process, when lowlands doctors examined the two thousand villagers for congenital health problems, making long reports about "gene pools" and "somatic types;" then the sociologists had come and spied on everyone, measuring both stability and resilience and adaptability of the existing village society; then the actual trainers, including men and women who had really walked on Mars (it was then that the headman realized that perhaps this was not just another lowlands taxation scheme), who identified skills needed on Mars, but which the villagers did not have. That had required sending off many young people for several years, while accepting strangers from other villages and even from the lowlands. The return of the young people had been the final proof of the authenticity of the project; up until then, the government had only stolen villagers for their wars, and few had returned.

The wind from the north reached the headman's nostrils. *Smell—that's the one thing that we will never duplicate. The odors were not at all unpleasant, just as the tastes of the crops were at first tolerable, and eventually delightful. But I'll never be able to close my eyes and not know where I am.*

Not everyone had been able to come. A tenth of them were refused passage for medical reasons, although age was not a primary factor. Many grandparents were among the emigrants, since the lowlands people knew that the rest would not leave without them. Some children stayed behind temporarily, but by that time, the headman was certain that he could trust the representatives directing their preparation. And he had been right. The children had joined them within a few years.

For twelve years the village had struggled with the new environment, which was not significantly different from the Andes in climate. There had been significant changes in their crops and their herd animals, and everyone had to spend more time in the new schoolhouse than they *would have preferred. That was another reason that his wife would not have imagined.* All of the literature, all of the technical books—they are in Quecha. *The young people no longer spoke with a lowlands accent, or sneered at their elders. Now they sought out their ancestors' poetry and wisdom, and already there were two or three among them improving upon it, molding their new experiences into the pattern of their ancient culture, and seeing Mars with a vision no European could ever understand. Strangely, their visitors found this new development even more exciting than their successes with their herds or their windmills.*

I'll never understand the lowlands people back home, or the lowlands people here, *the headman thought with a nod of his head. As he pulled out his string and quickly formed the variation cat's cradle his grandson had invented the other day, he also doubted if he would ever understand Mars.* But our children will, *he was certain, directing his thoughts to the sparsely populated hillside nearby where he knew there would soon be an additional family member.*

Mars, 2134 A.D.

Mars glitters in the human imagination even more brilliantly than it does in the night sky. Its reddish tint has excited human fancies for thousands of years, leading to the weaving of a complex pattern of myths about the "Red Planet." This red color reminded imaginative observers of blood, whose presence implied to them the activities of war and murder. Mars, the bleeding planet, came to be associated with violence on Earth.

Gradually, such mythology lost its grip on scientific thought. By the seventeenth century, most European philosophers were convinced that the universe was populated by many different species, each adapted to their home worlds. An image of a hypothetical Martian "man" was conjured up. Since Mars was smaller and colder than Earth, its inhabitants would naturally have been shaped by those conditions. And when it was later learned that the Martian air was thinner than that of Earth, this factor too was added to the equation. Mars "men" would be tall, furry, slight of build, but with gigantic chests containing powerful lungs.

Yet a countercurrent to this philosophy emerged at the beginning of this century. Instead of picturing a Martian culture shaped and molded by the harsh Martian conditions, Percival Lowell preferred to believe in an intelligent race of creatures who could build tremendous worldwide irrigation systems—the infamous canals—in order to counteract the effects of a deteriorating Martian climate.

Lowell's hypothetical Martians were engaging in planetary engineering. They were, in a word, terraforming their own planet. They were *not* adapting themselves to changing conditions, but were acting to counterbalance the change in the climatic conditions.

But it was all a mistake. There were no Martian canals, no Martian planetary engineering, and no Martians—at least not yet.

Fanciful scenarios of Martian life and climes filled the science fiction literature of the first half of this century. While astronomers measured out the meager resources of Mars, writers such as Edgar Rice Burroughs peopled the world of Barsoom with animals, men, and half-men. The theme that Mars was a dying planet was also popular, exploited by H. G. Well's *War of the Worlds* and carried on by dozens of other writers. Mars, it seemed, was spilling its precious lifeblood as it gradually slipped towards planetary death.

A vision of the real Mars, and of the potential for human engineering of Martian conditions, appeared in 1952 with the publication of Arthur Clarke's novel, *Sands of Mars*. Its imagery was for the most part hauntingly accurate; the later-disproved references to high mountains and little, furry Martian critters was understandable as artistic license.

Most important, human settlers on Clarke's Mars were preparing to terraform the planet. They faced both engineering and political problems. First was the challenge of warming the planet (which was to be accomplished by setting off a fusion reaction on one of Mars' tiny moons, creating a mini-sun) and creating oxygen (which was to be tackled with the help of especially designed vegetables called oxyfera plants). The costs of the project were, of course, large and subject to criticism from Earth politicians who thought it a mad scheme. In fact, according to Clarke's plot, the execution of "Project Dawn" was carried out by the colonists in secret, lest Earth officials forbid it.

Political difficulties also faced the Martian colonists of Isaac Asimov's short story *The Martian Way* (1953), who were faced with a cutoff of their precious water supply from Earth. In a desperate end run, a ragtag band of Martian space pilots voyaged to Saturn and diverted a miles-wide iceberg from Saturn's rings, laboriously nudging it back to Mars where its arrival broke the embargo. Although this story used the Saturnian water

The surface on Mars today. Although there is a thin atmosphere, it is so weak that these sand dunes have been essentially unchanged for hundreds of millions of years.

solely for life support for the colony, it demonstrated the possibility of the import of large quantities of volatiles from off-planet sources.

In the late 1950's, the 1960's, and the 1970's, the theme of terraforming Mars occurred again and again in science fiction. In general, the plots depended on some technological trick which the hero discovered, thanks to hints and a stacked deck previously arranged by the author. One of the most plausible schemes was *Birth of Fire* by Jerry Pournelle (1976), which showed an appreciation for the factors of planetary climate and the tools of would-be terraformers.

Pournelle's gimmick was to trigger the Martian volcanoes with atomic bombs, thus releasing pent up gases which would thicken the Martian atmosphere and help trap solar heat, thus leading to a general warming and a melting of other frozen volatiles. To make this story plausible, Pournelle had to claim that the Martian volcanoes had only been dormant for centuries, rather than billions of years.

The story required token skeptics who voiced their objections to such projects. One Pournelle character cried out: "We cannot destroy the ecology of an entire planet. To humans, perhaps, a breathable atmosphere on Mars is desirable, but to Mars it is no more than pollution. . . ." Despite such opposition, an intricate plot unfolds to the climax when the first jury-rigged A-bomb was set off, giving rise to a new world "born in fire and ice." The good guys won. The old Mars died.

So Mars' grip on the human imagination has continued. Old myths have faded, while new science fiction images have been conjured up to take their place. In particular, Mars has played a role in the development of the concept of terraforming.

However, terraformers will need better arguments at budget time and bigger tools than mere Pournellian A-bombs when it comes time to seriously attempt to modify the Martian climate. They will need to know much more about Mars than we know even today after Viking. Yet despite our ignorance, we have accomplished something: we can now begin to ask the right questions.

A key factor about Mars is that it is half the diameter of Earth and of Venus, and is 50% further from the Sun than Earth is. Calculating simply on physical laws and temporarily disregarding atmospheric effects, the surface temperature of Mars should be about −50°C compared to 80°C for Venus and about 0°C for Earth.

The atmospheres of Earth and Venus cause significant "greenhouse effects," boosting the Venusian temperature by 400°C and Earth's temperature by about 40°C. The Martian air is so thin that it contributes a mere 4° bonus, after accounting for its albedo which at 16% is only half that of Earth.

Some Martian characteristics are more convenient: the planet rotates in about the same period as Earth, and its axis is tilted at almost the same angle. However, its orbit is significantly more eccentric, swinging closer and further from the Sun during its year. It is closest to the Sun in the southern summer, which allows the southern polar ice cap to melt completely; the colder northern summer is not powerful enough to melt a residual northern polar ice cap.

The coincidence in axial tilt is only temporary, according to mathematicians. They have computed that, due to planetary perturbations, the Martian axial tilt can vary between 14° and 35° over a 50,000-year period. (Earth's axial tilt only varies a few degrees at most because of the balancing effect of the Moon.) This Martian variance has very important effects on the Martian climate and on a hypothetical artificial Earth-like atmosphere, as we shall see shortly.

Because the Martian air is so thin (about 6 millibars, equal to an altitude of 40 kilometers or 24 miles on Earth), it cannot carry a significant amount of heat, so the polar regions do not receive supplementary heating from warmer equatorial winds as on Earth. The thin air also means that surface temperatures are a measure only of solar flux, not of elevation from a mean Martian "sea level."

And the altitude variations of the Martian surface are greater than that of Earth, hinting that the Martian crust is thicker, colder, and hence less plastic than Earth's. The great volcano Nix Olympus rises some 25 kilometers (15 miles) above the surrounding plains; Arsia Mons is 20 kilometers (12 miles) high with a summit pressure of about 1 ½ millibars;

Pavonis Lacus is the tallest Martian mountain, with a summit air pressure of only 1 millibar.

The worldwide topography is also unbalanced, with the northern planes at a generally lower altitude than the southern highlands. As Mars geologist Thomas ("Tim") Mutch said, "If Mars had an ocean, it would cover the northern hemisphere, and the southern hemisphere would be a single vast continent." In a later scenario, we will consider a "Boreal Sea" and its effects on the entire planet.

One exception to this topographic pattern is the giant basin called Hellas, 1,600 kilometers (1,000 miles) in diameter, right in the heart of the southern highlands. In the middle of this basin (which is remarkably free of surface features) is the lowest point on Mars, where even now the air pressure is high enough for liquid water to remain stable.

The craters of Mars may be superficially similar to those of the Moon, Mercury, and other space-blasted worlds, but there are important differences. Smaller craters have about the same proportions as their lunar cousins, but larger craters are shallower. Fewer than half as many large basins occur than would be suspected from the number on the Moon and Mercury. Either the extrapolations are wrong, or some sort of powerful erosion was at work early in Mars' history. About half of the planet is surfaced by relatively young plains, especially in the low-lying northern hemisphere.

Climate History

Mars' climate, like that of Earth, has varied over the eons. The two most obvious indications of this fact are the channels which crisscross the surface and the *laminae* or layered deposits at the poles. Both suggest a much more hospitable former climate—a climate whose disappearance would-be terraformers intend to resurrect.

The striking photos of erosion on Mars, when studied carefully, have revealed some significant patterns. There are different types of erosive features and their distributions are quite different. Channels can be *outflow, runoff,* and *fretted;* other common erosive features are described as *gullies.*

Outflow channels are large features (up to 1,000 kilometers long and 100 kilometers wide) which begin full-born at a specific location. They are probably caused by the sudden release of subsurface water, perhaps water under artesian pressure which breaks through a layer of

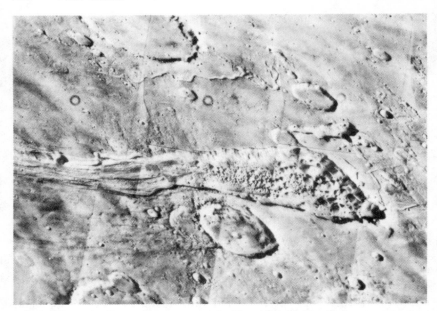

Other evidence for large subsurface water is this outflow crater which appears to show an area where subsurface water melted and ran out onto the surface.

permafrost, or perhaps ice melted by hot springs or meteorite impact. Since they are one time and short-lived events, they do not necessarily require the existence of a permanent thicker atmosphere.

Gullies are quite different. They are smaller and shorter by factors of 10 to 100. They occur in bunches, but they occur all over Mars except on the youngest surfaces, and they occur even on exterior and interior walls of impact craters, demonstrating that they were formed after the period of intense cratering. They are *not* found on the relatively young Tharsis Volcano, which itself is a few billion years old; this suggests that the factor which caused them (probably rainfall) has *not* been part of Mars climate for at least that long.

The runoff channels and the fretted channels seem to have formed over an extended period of time rather than all at once. They are rarer than the gullies, but they seem to be about the same age. Although they might be connected with the existence of a warmer, wetter Mars, their precise background is still obscure.

The second evidence of the existence of climate change on Mars is the *laminae*, the sedimentary layered deposits which lie under the polar regions. They are visible because in many areas something has eroded

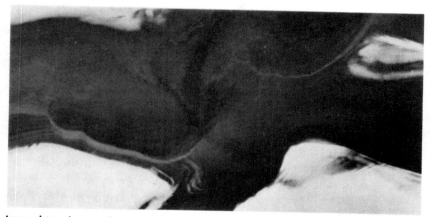

Layered terrain near the Martian north pole testifies to the existence of a much thicker Martian atmosphere billions of years ago. The processes which led to the formation of the layers and then to their random erosion are still not understood.

the surface down into the layers. These layered areas are among the youngest features on Mars, judging from the lack of significant meteor craters. The layers become thinner near the top, which suggests that the agency that caused them was losing its power in its later stages. The layers were presumably laid down and then randomly eroded by the wind. Since the erosion rate of the present atmosphere is now calculated at fractions of an inch per million years, a much thicker Martian atmosphere is definitely indicated.

Natural Climate Change of Mars

According to Milankovich's theory of orbital variations, ice ages on Earth were touched off by extreme cycles in the planet's orbit and axial tilt. The same factors are in effect for Mars. Earth, however, has a large nearby satellite whose gravitational forces tend to preclude any wild fluctuations. Mars has no such help, and its variations are correspondingly more severe.

Computing the combined effects of the Sun, nearby Jupiter, and the other planets on the motion of Mars, scientists have discovered several different cycles.[1] The eccentricity of the orbit of Mars around the Sun varies by 6% over a period of 95 thousand years and by another 10% over a period of 2 million years. Its axis of rotation bobs up and down from a tilt of 15° to 35°, which has a major effect on polar temperatures, which in turn drive the planet-wide climate. This bobbing axis also precesses or wobbles with a period of 175,000 years. All of these variations create changes in planetary heating, which probably lead to changes in the amount of the atmosphere frozen out and the amount free; the exact effect on Martian climate is still being investigated.

Changes in Martian climate can also be caused by changes in the composition of the atmosphere. James Pollack's excellent recent survey of "Climate Change on Terrestrial Planets" described the effect of possible early Mars atmospheres on Martian surface temperature, a computation which was complicated by the fact that the brightness of the Sun was 30% less when the planets were young, 3 billion years ago.[2]

According to Pollack, a Martian atmosphere of carbon dioxide alone would not produce a very efficient "greenhouse effect." In fact, a surface pressure of about 2 bars would be required for the globally and annually averaged surface temperatures to be above freezing. However, if the atmosphere were predominantly methane with admixtures of ammonia and water vapor, a very effective "greenhouse effect" would have been created which would have allowed liquid water to form even if the atmospheric pressure were as low as 0.15 bars. Planetary engineers should be aware of the various ways a "greenhouse effect" can be assembled, and the subsequent ways in which planet-wide biology can modulate this "greenhouse effect" via modification of the atmospheric composition.

Yet another factor in the changing climate of Mars is the off-balance weight load of the Tharsis volcanic formation on the spin axis of Mars. Dr. Owen Toon calculated in 1977 that the formation of this massive volcanic shield was sufficiently destabilizing to reduce the average axial tilt of the planet from 34° to the present 25°, thus lessening the average solar heating of the poles and allowing most of the free air to freeze out into the ice caps. Previously, Toon speculates, there should have been frequent occurrences of ice melting on the surface even without a "greenhouse effect."[3]

The lessons of these natural climate-changing factors can help terraformers who wish one day to artificially duplicate natural effects. Atmospheric changes can be induced by the import of volatile materials in asteroids or by the retrieval of native Mars volatiles locked away in permafrost. Solar input can be redistributed and enhanced with orbiting space mirrors. Mass distribution on the surface can be manipulated by the creation of artificial *mascons,* mass concentrations left by the impact

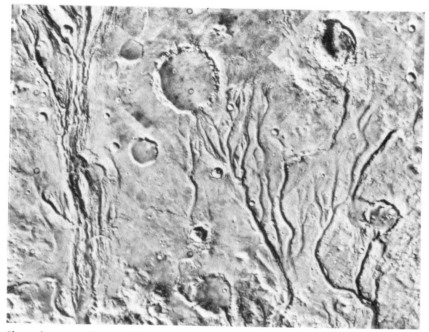

Channels apparently carved by running water are plentiful on the Martian surface, hinting at the existence of a former warm climate. Terraformers may think of themselves restoring the planet rather than raping it.

of large stony-iron asteroids. All of these changes, which have occurred by accident in the billions of years of Mars history, can serve as screenplays for the re-creation of selected scenes; if the encores are choreographed properly, the desired Earth-like Mars might eventually spring forth.

Water on Mars

Leaving aside the clear evidence for episodic (and very ancient) flooding in the channels, there is an interesting body of evidence which suggests that much of Mars' surface is underlaid with thick beds of frozen water or permafrost. This is not based on planetary formation arguments about what *should* have collected on Mars, but is based on real physical features observed on the Martian surface by space probes.

A unique feature of Martian impact structures is the occurrence of *lobate* craters, more descriptively called "splosh craters." The rim ma-

Vast reservoirs of permafrost may underlie the Martian surface, as hinted by this very *muddy* looking crater. Such craters, called "splosh" or "lobate" craters, are very common across the face of Mars.

terial appears to have oozed away from the impact rather than to have been merely thrown. Several explanations have been proposed for such craters: they represent the impact of an object into subsurface water; they are caused by the entrapment of atmospheric gases by escaping debris during the impact; they are caused by the impact of water-bearing asteroids; they are a little-understood result of wind erosion of ordinary impact craters. Also, a theory states that they may have been caused by the later melting of subsurface water by overlaying hot *ejecta* from an impact. Yet for each explanation, a counter-example can be found to eliminate it.

Statistical evidence is informative: such craters do not preferentially occur on any particular geologic region, nor is there any correlation with elevation. There is, however, an apparent variance as a function of *latitude*, which is a distribution consistent with possible atmospheric volatile distribution in the past. This global arrangement of "splosh craters" suggests that they are evidence for the thermal redistribution of whatever feature that causes them. The most obvious feature is subsurface water ice.

Additional geologic activities may show distinguishable traces of the effects of water or of ice. Dr. Carleton C. Allen of the University of Arizona announced in 1979 that he has identified numerous Martian analogs for Earth features found in Iceland—features created by the action of volcanic eruptions through glaciers.[4]

These features are called *table mountains, moberg ridges,* and *pseudocraters.* The table mountains are caused by lava eruptions under ice; they have distinctive shapes, including steep sides and flat tops. Mobergs are products of subglacial fissure lava eruptions along a line. Pseudocraters are formed when lava flows over water-saturated ground and the water bursts violently into steam.

Such features, Allen reports, are common in the northern plains of Mars, specifically in Amazonis, Chryse, Utopia, and, most notably, Acidalia Planitia. Since they show where ice *used* to be, they suggest to Allen that the northern hemisphere of Mars was once covered with up to a kilometer (more than one-half mile) of water. Since these features are not characteristic of the southern highlands, those regions may still have that much water locked up as buried glaciers or even a buried ocean.

Numerous cracks in the Martian surface, as observed from orbiting satellites, are similar to terrestrial permafrost networks. Radar returns from Mars in 1976 also suggested the present existence of thick permafrost. A fairly consistent picture has emerged: Mars apparently still has a large volume of water and other volatiles tucked away in deep freeze under the sands.

One Mars expert who is convinced of this is Dr. Thomas Mutch, formerly NASA's chief space scientist. In *The Geology of Mars,* he wrote: "There might be a large amount of unseen volatiles in the planet's surface and subsurface material that we have not considered. A substantial if not predominant quantity of. . . . volatiles may exist beneath the Martian surface at all latitudes, both as ground ice and absorbed on mineral grains."[5] The visible ice in the polar cap may thus, some scientists believe, represent only a small fraction of the total available Martian water.

The polar caps of Mars are still not well understood, even though they are important to schemes of Martian terraforming. The caps are not symmetrical from north to south, either, as a result of the great eccentricity of Mars' orbit.

The northern cap, which extends to 60° N at a maximum, shrinks to a permanent residue during northern summer. This residue lies north of 82° N and is surrounded by a bizarre series of terraces and other erosional features. The volume of the residual cap is still unknown; it could be anywhere from a meter to a kilometer thick. It is almost certainly

The permanent north polar cap of Mars almost certainly is water ice. Far more water ice is probably located under the planet's surface in the form of permafrost. The solid white area toward the top (north) is ice—probably both frozen carbon dioxide and water ice. The dark bands, regions devoid of ice, spiral in toward the cap's center. The reason so little ice occurs in these bands is uncertain but may be related to winds blowing away from the center of the cap.

water ice, while the variable sections of both polar caps are carbon dioxide ice ("dry ice").

The southern cap extends halfway to the equator in winter, since the southern winter is longer than the northern one. It melts completely in the summer, and its effect on atmospheric pressure can be used to compute its volume with high accuracy. The model thus created has an average depth of 23 centimeters (10 inches), with maximum depths near the pole of about 1 meter (a few feet) at most. And geographically, the land on which the southern cap rests is several kilometers higher than the northern cap, another factor in the apparent asymmetry.

History of Martian Volatiles

We have now outlined the evidence that large deposits of water lie frozen beneath the Martian surface. To understand where this water (and other volatiles) might have come from, we need to look at the theoretical models of planetary outgassing which, presumably, led to the current inventory of volatiles on Mars.

From the argon isotope ratios measured by Viking, scientists have concluded that Mars outgassed about 100 times *less* efficiently than did Earth. This is not particularly surprising since the smaller size of the planet had already suggested that it would not have generated nearly as much

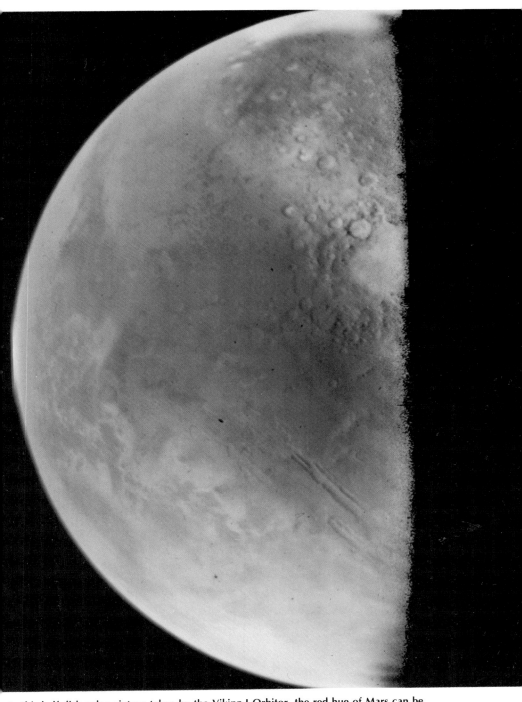
In this half-disk color picture taken by the Viking I Orbitor, the red hue of Mars can be seen with varying degrees of brightness suggesting planet compositional differences.

The transformation of the moon. Above, the Mare Imbrium as it is today. Below, the same region, after the addition of a small amount of volatile asteroidal compounds and water, is shown to be basically the same in appearance. Designing the appropriate atmosphere, hydrosphere, and biosphere is a complex but feasible project and could possibly begin by the end of the next century.

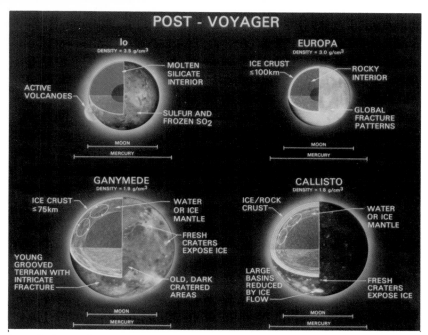

Once the inner planets of the Solar System have been started toward their terraforming future, the Galilean satellites of Jupiter are logically next in line. Io has a convenient internal heat source but needs an atmosphere; Europa has an adequate frozen ocean but needs heat; Ganymede and Callisto have vast quantities of water which might be useful in terraforming Mercury.

This explosion on Io, recognizable by the light green spray to the left, gives evidence of an internal heat source that would make this Jovian moon a prime resource for terraforming.

In this conception of a terraformed Mars, artificial activities have melted the Martian glaciers and dismantled one of its small moons for use as surface dirt. The "Boreal Sea" in the northern hemisphere supports hurricanes and warm currents; vast prairies cover the equatorial regions.

In this fanciful view of a terraformed Venus, a cloud-shrouded equatorial ocean girds a "new" world whose polar continents have climates akin to Earth's South Sea Islands.

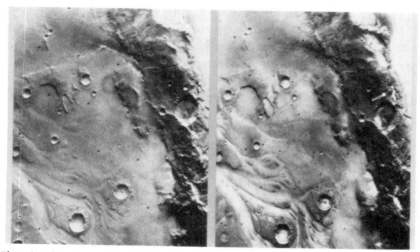

There is still water on Mars; in fact, the atmospheric humidity is nearly 100% (which, in Mars's thin air, is not saying much). Here, ground fog is spotted by cameras on the Viking 1 mother-ship high above the planet. The fog appeared soon after dawn but was gone within half an hour.

heat as Earth; consequently, the volumes of gas released on Earth were not matched on Mars. They are still locked deep in the interior and are likely to remain there forever.

Based on Viking data, planetary atmospheric specialist Joel Levine of NASA's Langley Research Center has estimated in 1978 that Mars had outgassed about 2 bars worth of water (averaged over the surface of Mars, that comes to 50 meters or 190 feet) with perhaps six times more or ten times less.[6] Although Mars is today losing water to space at a rate of 60,000 gallons per day, that is an inconsequential amount even for that parched planet. It would, however, account for about 3 meters (10 feet) of water lost over the lifetime of Mars.

Where is the rest of the water? Levine theorizes that some could be physically absorbed in the rocky rubble of the soil, but that many times that amount could be in one or both of two hiding places: the water could be chemically bound in clay and other minerals, or it could exist in the form of ice and permafrost.

So if the photographic data suggests extensive permafrost, Levine's results confirm that such vast amounts of water would be quite consistent with what we have learned about the evolution of the Martian atmosphere.

Similar results are available for carbon dioxide and nitrogen. Whereas only 7 millibars of carbon dioxide remain free, Levine's analysis suggests that between 37 and 370 millibars were outgassed, disappearing into chemical combinations with the soil. Other data suggests that up to 30 millibars of nitrogen may have been released, although only 0.2 millibar remains free. The present level on Earth is 750 millibars.

Melting the Martian Ice

An interesting version of a pre-Viking Mars terraforming scheme was published in 1973. Cornell University scientists Joseph A. Burns and Martin Harwit, in their article *Towards a More Habitable Mars—or—the Coming Martian Spring*, described a theory by Carl Sagan that Mars only occasionally has liquid water in its atmosphere, but generally has the water locked up in either polar cap.[7] The Burns-Harwit plan would "freeze" the Martian climate in the status of that "Martian spring" when the water remained free in the atmosphere.

Sagan's pre-Viking suggestion was that the 50,000-year cycle of Martian orbital and spin axis variations results in periodic cold spells when one pole is leaning away from the Sun during summer. This resulted in sufficient cold to maintain all Martian water frozen solid year round in that polar cap. As the spin axis of Mars shifts around, the frozen pole eventually is no longer facing away from the summer Sun and the water can melt. Soon, however, the other pole swings around into the appropriate direction and the temporarily free water condenses there and freezes into a new ice cap.

Burns and Harwit suggest that this "wobble" of the Martian pole be stopped at a point in the cycle when neither pole is acting like a deep freeze. This would leave the water permanently free to work its biological wonders on the planet. To do this, they suggest using a large object in orbit around Mars as a counterweight to the slow torque of the Sun, so that the relative angles of the Martian orbit and axis remain constant.

That's a nice theory and the mathematics are rather elegant. But in practice, where is such a counterweight? The inner moon Phobos is suggested first, and an appropriate orbit is computed. It turns out that moving Phobos into such an orbit would require 100% efficient conversion of all sunlight falling on the moonlet for 10,000 years, or an equivalent sum of energy, poured in more rapidly. But Phobos fails the needs of the project, since it would be quickly torn out of the necessary orbit by the gravitational perturbations of Mars itself.

Another candidate is a nearby asteroid that could be placed in a special retrograde orbit of Mars. The torquing force can actually come from a single object, a group of objects, or a new artificial ring around Mars. With some care, the desired effects would be attained.

Burns and Harwit suggest that their proposal is environmentally sound: "Although there is always something a little repugnant about man pushing his own interests and fixing nature, we believe that—of all possible ways to prolong the spring—the above scheme would do the least to directly damage Mars. This occurs because foreign matter is never introduced onto Mars' surface nor into its atmosphere; everything is done externally. The proposal is, perhaps, a fantastic one to contemporary minds. However, it seems to us that the required technology will not be wanting if man is alive 10,000 years from now."

The basis for the "Burns-Harwit Maneuver," recall, is the assumption that Mars is still going through its cycles of climate change and that the present era is one of "long winter." However, more recent data analysis has convinced Mars geologists that the planet is no longer going through any periodic changes, and that the "long spring" for which Burns and Harwit were waiting will never come if we leave nature to take its course.

However, the "Burns-Harwit Maneuver" will still be useful. One important problem in the long term habitability of Mars is the wide variations in its orbit and axis—variations which can be severe enough to destroy a fragile planetary biosphere. In the far future, once Mars is terraformed and inhabited, its civilization may decide to forestall such wide and dangerous variations through the same technique which has been so successful on Earth: the presence of a nearby, large satellite. The emplacement of such a satellite (perhaps Ganymede, stolen from Jupiter) will fulfill the predictions made by Burns and Harwit, although for a different underlying purpose. People, not blind natural cycles, will have brought the "Martian spring," and people will act to preserve it.

NASA's Martian Ecosynthesis

Just melting the ice is only the beginning in rebuilding Mars. Even if the air were thick and the winds warm, the atmosphere of carbon dioxide, water vapor, and other imported trace gases such as methane and ammonia would hardly be the new version of Earth which terraformers desire. The composition of the atmosphere must be changed, and biology provides the most promising tools: lichen, algae, methanogens, and other microorganisms.

164   New Earths

Biological processes were at the heart of the NASA-Ames study on terraforming Mars, conducted in 1975. The question which was asked was "whether Mars is a habitable planet or *can be made into one*."[8] Based primarily on pre-Viking data, the study provided the first in-depth look at what it would take to allow terrestrial organisms to take root on Mars and begin the task of converting the atmosphere. They summarized their results as follows:

> The creation of an adequate oxygen and ozone-containing atmosphere on Mars may be feasible through the use of photosynthetic organisms. The time needed to generate such an atmosphere, however, might be several millions of years. This period might be drastically reduced by (1) the synthesis of novel, Mars-adapted, oxygen producing photosynthetic strains by techniques of genetic engineering, and (2) modifying the present Martian climate by techniques of planetary engineering. Such climatic modification would rely upon the melting of the Martian polar caps and concommitant advective and greenhouse heating effects. Melting the polar caps, however, would require the investment of very large amounts of energy over a relatively long period.

These ideas may seem staggering even for the ingenious group at NASA. But such optimistic ideas were evident even in the title of the booklet which described the results of the project: "On the Habitability of Mars—An Approach to Planetary Ecosynthesis." Unfettered by budgets or by ordinary engineering complications, the would-be terraformers of Mars attacked the problem with pure logic.

And what a problem it was! Seeding Mars presented many difficulties: "The predominant environmental factors facing organisms seeded on the surface of Mars . . . include: mean temperatures below freezing, large day-night temperature variations, strong ultraviolet irradiation, high dessicating conditions, and frequent abrasive dust storms [although Viking was to find that the force of the dust storms had been grossly overrated]."

After examining available Earth plants which might survive on Mars, and postulating the creation of a special *Marsophile* (Mars loving) breed, NASA scientists were still faced with a dreadfully slow process. "Temperatures and water loss would limit photosynthesis to only 3 to 5 hours per day," the report concluded. "Allowing for a 25% coverage of the surface of Mars, blue-green algae could generate an amount of oxygen equivalent to the present amount of carbon dioxide in the Martian at-

BIOLOGICAL AND GROWTH CHARACTERISTICS OF SOME TERRESTRIAL ORGANISMS AND OF AN IDEAL MARTIAN ORGANISM

Organism	Requires oxygen	Extreme resistance to ultraviolet radiation	Extreme resistance to drying	Growth rate	Growth habitat
Green algae	Yes	No	No	Fast (hr)	Soil (surface and subsurface), Snow (surface), water
Lichen	Yes	Yes	Yes	Very slow (yr)	Surfaces (rock, tree)
Moss	Yes	No	No	Slow (wk)	Moist surfaces
Blue-green algae	No	No	Yes	Fast (hr)	Soil (surface and subsurface), water
Ideal Martian organism	No	Yes	Yes	Very fast (min)	Soil (surface and subsurface), water

In 1975 NASA studied ways to use Earth plants for the conversion of the Martian atmosphere. Candidates were rated according to varying characteristics useful for survival on present-day Mars.

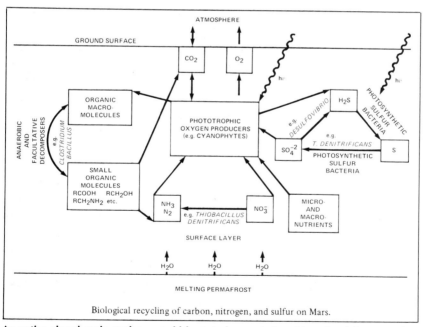

Biological recycling of carbon, nitrogen, and sulfur on Mars.

An entire closed-cycle ecology would have to be created on Mars, involving a number of different types of micro-organisms. So far, nobody on Earth knows how to create computer models to predict the behavior of such systems as they are being established.

mosphere (approximately 5 millibars) in 7000 years. To produce an amount equivalent to the minimum necessary for human breathing (approximately 100 millibars) would take 140,000 years. The lichens would take approximately 10 times longer."

With these unacceptable numbers in front of them, the terraformers looked for short cuts—for clever, roundabout ways to overcome these limitations. They found them.

"Many of the environmental factors which are unfavorable for the establishment of life on Mars could be eased by modifying the planet's temperature," the report stated matter-of-factly. "The surface temperatures of Mars could be increased, and the day-night temperature variations reduced if a means could be found for increasing the atmospheric mass by vaporizing the polar caps."

Additional gases in the atmosphere could help hold heat in. However, the "greenhouse effect" from carbon dioxide alone is small; the addition of enough carbon dioxide to increase surface pressure a hundredfold would still only raise the surface temperature 7°C. Water provides a much more powerful "greenhouse effect"; enough water to increase the present atmosphere only 10% would provide an incredible temperature bonus of 10°C.

Additional gases in the atmosphere would also help carry heat from the equator to the poles. On Earth, the atmosphere helps overcome the polar energy deficit, and this prevents those regions from freezing out all the moisture in the atmosphere. The term for this effect was *advective heating* (heat carried horizontally) and it plays a crucial role in keeping the Earth's biosphere habitable. It could do the same on Mars if the air were made thick enough.

"In brief," the NASA report explained, "there are two stable climatic regimes possible on Mars, given a carbon dioxide polar cap during the winter season. The present climate represents one of the stable regimes. The second stable regime exists at a polar surface winter temperature of $-80°C$ ($-100°F$) and at a surface air pressure of approximately 1 bar."

A recipe for converting the planet to the more Earth-like regime was provided by the NASA team: "To attain this high temperature regime the surface pressure must be increased ten times. This might be brought about in a number of ways: by volatilizing gases bound in the *regolith* [italics added], by importing an atmosphere from outside the planet, or by increasing the effective solar flux over the polar cap by 20%."

The last technique could be accomplished by using mirrors in space

Mars—A First Look 167

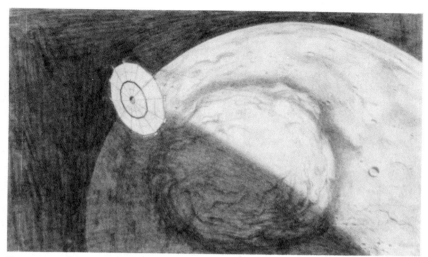

One approach to terraforming Mars is to increase the amount of sunlight reaching the poles. Here, giant mirrors reflect solar heat onto a polar ice cap.

to concentrate additional solar energy, or, perhaps more efficiently, by reducing the albedo of the ice from its present value of 77%. If the albedo were just reduced to say 73%, perhaps by the deposition of a layer of darker material atop the ice, enough sunlight would be absorbed in one century to trigger the release of a larger atmosphere which would set up a "greenhouse effect" and advective heating.

Another option involves the creation of special Mars-designed biota. The report continues:

> It is interesting to speculate upon the possibility of creating novel species of photosynthetic organisms far better adapted to growth in the present or modified Martian environment: in effect, transforming currently available "best fit" organisms into "ideal" organisms. Such genetic engineering is possible, utilizing methods of gene manipulation currently known or under development.
>
> Research on bacteria and their viruses has yielded powerful tools for the manipulation of the genetic apparatus of cells. Genes determine the protein enzymes of cells and these (to a great extent) determine the physical characteristics of cells. Thus if a cell has a certain characteristic, for example, capability of rapid repair of ultraviolet radiation damage, it is due to the presence of a certain gene or of several genes. A cell which lacks such a rapid repair mechanism would lack this gene or genes. Today there are techniques for moving genes from donor organisms and inserting them into recipient or-

ganisms. Several of these techniques, that is, sexual recombination and DNA transformation, have been successfully applied in species of blue-green algae. Thus while no single species of blue-green algae may have all those characteristics defining an ideal Martian organism, they well might be found in several different species of algae—all utilizable as gene donors. Indeed, in principle, the entire gene pool of the Earth might be available for the construction of an ideally adapted oxygen-producing photosynthetic Martian organisms. Continued advances in our understanding of gene structure and of methods for the synthesis of genes may lead to a potential for creating novel genes. These synthetic genes could then be integrated into a recipient "best fit" organism.

The Earth organism most attractive to the NASA Mars terraformers was the cyanophytes or blue-green algae. (Actually some plant physiologists insist they are really a species of bacteria.) This is a group of related species which may be particularly well suited to genetic tampering to increase their tolerance (even enjoyment) of Martian conditions. As discussed previously, such plants could be powerful tools for rebuilding planetary atmospheres.

NASA concluded that things looked pretty good. However, many proposals depended on untested assumptions. "Factors as yet undetermined might make dissemination of terrestrial organisms on Mars unsuccessful," NASA warned. The lack of subsurface ice fields close enough to the surface to be easily melted, a deficiency of elements essential to terrestrial life, or the presence of toxic compounds would further encumber attempts to establish terrestrial organisms on the planet." Until the Vikings landed, NASA could only guess.

Oasis Craters of Mars

According to the NASA study on terraforming Mars, a major problem in using biological agents to convert carbon dioxide into oxygen and carbon compounds is that the air pressure now on Mars is much too thin to support significant activities. The NASA study suggested planetary engineering to raise the air pressure all over Mars.

That may be too drastic a measure for a first step. We may be satisfied with raising the air pressure over just a small portion of the Martian surface.

There are a number of ways of doing this. Large, classical domed cities would form one approach, with the thicker air held down by means

Deep craters dug by the guided impacts of water-rich asteroids may provide "oases" on Mars as the first footholds of terrestrial life forms.

of walls. In general, any technique which piles more air over a given spot would be adequate—and there is just such a technique. The answer is to find *or create* low altitude regions on the planet.

Hellas is the lowest region on Mars which occurs naturally so we will have to dig a deeper hole. To get a pressure of 500 millibars (half that of Earth's surface), a hole 40 kilometers (25 miles) deep would be needed.

One deep shaft would not be of much use, since sunlight would rarely reach the bottom where we would place our open air gardens. But a region of several thousand square kilometers with sloping sides would be ideal.

Where would the energy come from? Excavating by tractor would be inordinately slow. Excavating by nuclear explosive would require thousands of the biggest thermonuclear bombs around. So where do we get the energy?

It's already on its way. The Saturnian icebergs with their precious loads of water and other volatiles could be impacting Mars with velocities above 65 kilometers (40 miles) per second. Craters 40 kilometers (25 miles) deep could be dug by *iceteroids* a few kilometers across; an object the size of Phoebe, kidnapped from its distant Saturnian orbit and turned towards Mars, could be split into enough projectiles to dig dozens of deep craters.

That, at least, is the conclusion of Dr. A.W.G. Kunze of the University of Akron's Department of Geology. For example, he calculates that "an asteroid 67 kilometers in diameter with a density of 3 g/cc [grams per cubic centimeter] impacting at escape velocity—5.1 km/sec—would excavate a 41 km deep crater—deep enough for 500 mb [millibars] atmospheric pressure."[9] That's twice as much as is needed for a human being to walk around in the open, assuming he or she wore warm clothing and an oxygen mask.

Furthermore, the added mass of Saturnian volatiles, together with other gases sure to be blasted out of the Martian subsurface, would increase the Martian atmospheric mass severalfold and would result in requiring a correspondingly shallower crater to obtain the same desired pressures. The depth of the needed craters may be only 8 to 15 kilometers (5 to 10 miles) in the final analysis.

The impact would also heat the surrounding rocks while vaporizing the ice and sending it into the atmosphere as clouds, soon to condense and fall as rain or snow. The resulting crater would thus have a significant artificial warmth as well as a thicker atmosphere—ideal conditions in which to grow the vegetation which is to transform the Martian atmosphere. It is therefore not necessary to cover the entire Martian surface with hardy, inefficient plants; much more pleasant conditions on a small fraction of the surface (say, a few tenths of a percent) could achieve the same ends. The actual function of these atmosphere-concentrating depressions may provide a less clumsy appellation than *oasis craters*. We could call them "air holes."

The different oasis craters would, at least initially, be biologically isolated from each other, with killing ultraviolet rays still bathing the highlands between the planted regions. This quarantine would allow time for experimentation and parallel developments in different craters, in order to prove out the theories developed on Earth and to find the best adaptations to the true Martian environment.

The craters would gather water, too, first in central crater lakes and later (over the decades as the atmosphere thickens), spreading out to engulf and fill the entire oasis. As this occurs, the highland regions will also become inhabited by spreading plant life. In the end, the craters will be abandoned to the new Martian lakes and oceans.

In fact, it may be most appropriate to develop life in these bodies of water even before much effort is given to life on the dry land. The water would provide thermal stability and even better protection against dust storms and ultraviolet rays. As on Earth, the water would be a habitable biosphere long before the dry land would be fit for extensive settlement.

Summing Up

As the Vikings approached Mars in the summer of 1976, the planet appeared to be an extremely attractive target for planetary engineering. Some areas of ignorance were glossed over with reasonable guesses. The Mariner-9 photographs (1971–1973) had provided plenty of evidence for estimating the history of the planet's climate and for attempting to inventory the planet's water supply.

Several suggested scenarios for rebuilding the planet were already widely publicized. Science fiction had pioneered the topic, but scientific speculations had not been far behind, and NASA had brought up the rear with an officially sanctioned study project. Techniques suggested included importing ice from Saturn, adding a large moon to stabilize the planet's orbit, and dropping asteroids onto the surface to dig deep "air holes" which would contain quite mild climates.

These imaginative proposals would soon be put to the test when the new data from the Vikings appeared. Mars had already revealed itself to be a surprisingly complex world. Would the Mars discovered by the Vikings be easier or harder to terraform?

MARS—AN INTERLUDE

Mars is so very tempting! Each time you spot something that might be insuperable, you work on it and sooner or later a way around it appears. Mars is, at this time, out of reach, but *not very* out of reach.

> Melvin Averner, NASA Mars Ecosynthesis Study (1979)

We *want* Mars to be like the Earth. There is a very deep-seated desire to find another place where we can make another start, that somehow could be habitable. . . . It has been very, very hard to face up to the facts, which have been emerging for some time, that indicate it really isn't that way, that it is just wishful thinking.

> Bruce Murray

They had a house of crystal pillars on the planet Mars by the edge of an empty sea.

> Ray Bradbury, *The Martian Chronicles,* 1950

> Hesper—Venus—were we native to that splendour, or in Mars, we should see the Globe we groan in fairest of their evening stars.
>
> Alfred, Lord Tennyson

8

Mars—A Closer Look

Over a space of thirty years, notions of terraforming Mars matured markedly, as exemplified in the works of Arthur C. Clarke. Between the *Sands of Mars* (written in the late 1940's) and *Fountains of Paradise* (written in the mid 1970's), the visionary space writer did more than just change the code name from "Project Dawn" to "Project Eos" (Greek for "dawn"). He moved from some very good guesses later demolished by reality, to some very good scenarios based on our latest view of Martian realities.

A minor character in Clarke's latest (and reportedly last) novel is used as a mouthpiece for terraforming theories: "If we could thaw out all that water and carbon dioxide ice, several things would happen. The atmospheric density would increase, until men could work in the open without spacesuits. At a later stage, the air might even be made breathable. There would be running water, small seas, and, above all, vegetation—the beginnings of a carefully planned biota. In a couple of cen-

Mars—A Closer Look 175

Mars on its way to becoming a new Earth, as carbonaceous material from Phobos is propelled down onto the polar caps by giant "mass drivers."

turies, Mars could be another garden of Eden. It's the only planet in the Solar System we can transform with known technology. Venus may always be too hot. . . ."[1]

The connection of this subplot with the main theme of the novel, the construction of a space elevator between Earth's surface and a geosynchronous (*synchronized with Earth*) satellite 20,000 miles in space, is that a Martian version of such a space elevator would be needed to build the tools needed to terraform Mars: "The only practical way to heat up Mars," continues the explanation, "is by solar mirrors, hundreds of kilometers across. And we'll need them *permanently*—first to melt the icecaps, and later to maintain a comfortable temperature." The mirrors would be made out of sodium, plentiful on the surface of Mars but presumably rare in space; hence the need for an economical way to transport millions of tons of material off the Martian surface into space.

Viking Weathermen on Mars

The emplacement of two meterological stations on the Martian surface, together with the instrumented orbiting mother ships, gave scientists a new view of the Martian climate. Some guesses were confirmed, some new readings were gathered, and some eyebrows were raised.[2]

When the Viking probes landed on Mars in 1976 they began direct human modification of that planet's surface (the hole was dug by a mechanical arm).

The composition of the atmosphere was about as expected: 96% carbon dioxide, almost 3% nitrogen (some good news for the biologists, but not enough), almost 2% argon, and traces of oxygen (0.13%), carbon monoxide (0.07%), and water vapor (0.03%). Nor are the biological experiments, however exciting and perplexing, of direct relevance at the moment. We do want to know if there is life on Mars, and we do want to examine the bizarre surface chemistry of the planet—but that can wait. What is of immediate interest is the weather.

The dynamics of the Martian atmosphere help drive the climate changes which do occur. By studying these interactions, would-be terraformers may discover a fulcrum for a technological or biological lever.

Major differences between Earth and Mars meteorology is due to the major roles played by carbon dioxide and by atmospheric dust on Mars, which affects the way in which heat is exchanged between the atmosphere, space, and the ground.

Contrary to expectations, the Martian air was found to be capable of carrying a small amount of heat by advection. Viking found an 8°C

Mars—A Closer Look   177

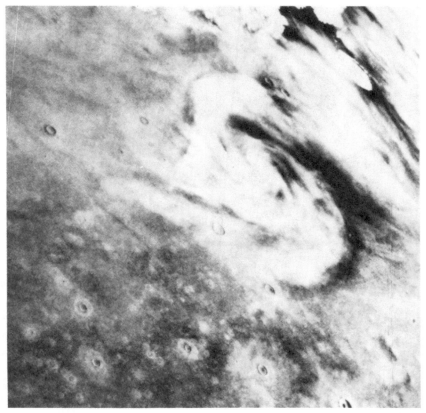

The Martian atmosphere is already thick enough to form hurricanes hundreds of miles across.

difference between northerly and southerly winds. Also, the Viking orbiters found that the temperature of the dark, southern winter polar cap is measurably warmer than it should be, and advective heating from the equator is the only possible mechanism.

Dust in the Martian atmosphere seems to have some of the heat-transporting characteristics of water vapor in Earth's atmosphere, but unlike Earth's, it never gets to form condensation nuclei for raindrops. The feedback system of the Great Martian Dust Storms seems to control climate on a worldwide scale and shows a few parallels with the water-atmosphere cycle on Earth. These global dust storms occur during northern winter (but not during southern winter), when dust in the air absorbs

178   New Earths

Dust storms are the most prominent meteorological feature on Mars today. This view shows the whole planet shrouded in dust clouds.

heat and warms the air, setting it into motion.

And there is a lot of dust in the air, giving the Martian sky its characteristic red appearance. In fact, there is more dust in the Martian air than there is in Earth's air: on the average there is about ten times as much as is found over a large American city on a smoggy day. That's because there is no rain to wash out the dust, so it stays and accumulates.

This heat absorption can raise atmospheric temperatures by as much as 25°C in a single day. Wind is generated, which picks up more dust, which catches more sunlight, until the whole planet is veiled in dust. On Earth, energy builds up in hurricanes by means of the action of water condensation, but once it has deposited its energy into the wind system, the water rains out, keeping the hurricanes "local." Martian dust, however, continues to add heat (captured solar heat rather than latent heat of vaporization) into the wind systems for months at a time.

In another aspect of meteorology, Mars is still opposite of Earth. In certain seasons, the optical depth of the air in the morning is higher than in the afternoon, because of the presence of morning ground fog which suppresses dust. On Earth, fog causes a reduction in air transparency; on Mars, just the opposite happens!

One Martian bogeyman was dispelled—the notion that killer sandstorms scour the surface, abrading exposed life forms to powder. Instead, "Meteorological observations suggest that about 1.6 meters [5 feet] above the surface of Mars wind speeds are not great (2–7 meters/sec), although previous observational history suggest[ed] massive, planet-wide dust storms that would be harmful to exposed organisms."[3] The storms exist, but not at the surface; it was a pleasant surprise.

The Question of Salt

One key question about the proposed Martian lakes and oceans is their salinity. How much salt will be dissolved in these waters? Will there be concentrations of other materials poisonous to Earth life, such as some of the heavy metals like mercury? These questions remain open, but there are some hints that this topic could be a major problem—a looming roadblock on the route to a habitable Mars.

As discussed at the 1979 Houston terraforming colloquium by Dr. Benton Clark of the Martin-Marietta Company (builders of the Viking probes), the Viking instruments measured the elemental and proportional characteristics of the Martian soil but did *not* give mineralogical analyses. This information had to be extrapolated from the raw Viking data.

Based on all the evidence at hand, the Martian soil seems to be predominantly (79%) *montmorillonite* clay, a combination of silicates and metals. About 13% of the soil is in the form of sulfates of sodium, magnesium, and other metals. Calcium carbonate makes up about 7% of the soil. There also appear to be some chlorides.

There may be too much salt in the surface material to allow easy plant growth. Dr. C.W. Snyder of the Jet Propulsion Laboratory told the Second Mars Colloquium in January 1979 that bromine levels, measured at 100 parts per million, were ten times higher than had been expected. "Contrary to the case on the moon," Snyder warned, "Martian 'soil' would be too salty to support agriculture."[4] Other Viking analysts, however, are not nearly as pessimistic, and believe Snyder's conclusion is unjustified: different interpretations of the elemental analyses do not support the existence of salty compounds.

On Earth, though, there is enough salt dissolved in the oceans to

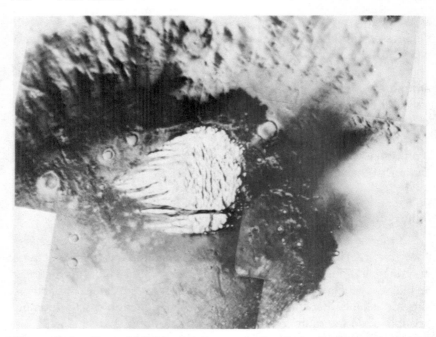

The so-called "white rock" inside a Martian crater raised hopes that unique deposits of materials—perhaps salts—had been discovered. In fact the rock is not white at all but only looks that way due to computer enhancement: it is actually dark grey, and is probably wind-eroded sediments of the native bedrock underlying the crater.

provide a layer more than 100 meters thick across the entire planet, or 400 meters thick on the dry land. Contrary to some common misconceptions, the saltiness of the ocean has not increased significantly since it first formed; as some salt has washed into the sea, a nearly equal amount has been locked up in the sediments forming on the ocean bottoms. Now, without open water, rainfall, and runoff, that salt would still be intermixed with the soil. The question is, how long did it take the rain to wash this salt out of the soil: a century, several millennia, a million years, or more? This question is important in trying to predict the effect of postulated Martian rains on hypothetical Martian soil salt and the salinity of any oceans which may be created on Mars.

No Tooth Fairy on Mars

One of the surprises of the Viking landers in 1976 was the observation that Martian soil, when moistened, gave off great amounts of free oxygen. This seems to have been a chemical rather than a biological process, and it conjured up visions of an unexpected bonus to the proposed watering of Mars for terraforming: the oxygen atmosphere might spring forth without tedious biological processes.

But these "great amounts" were not nearly as great in absolute terms—only in terms of what was expected. NASA's reconsideration of Martian eco-synthesis gave the death blow to that "free oxygen" concept: "Even if oxygen-containing material extended to depths of tens of meters," wrote Robert MacElroy in 1979, "oxygen contribution to the atmosphere, if the soil was all wetted, would be small, amounting at most to 1 to 2 percent of the mass of the present atmosphere."[5]

NASA Takes a Second Look

While it took Arthur Clarke thirty years to recycle his plan for terraforming Mars, it took NASA only four years. Based on Viking data analysis, no "show-stoppers" were discovered. However, the conclusion of this second look was markedly more pessimistic than the first study.

The planetary engineering program originally put forth by the 1975 study, "is insufficiently supported by either fact or calculation," wrote study director Dr. Robert MacElroy in 1979. "On the other hand, neither the physical nor biological assumptions have been demonstrated to be wrong, although several pieces of data from the Viking experiments strongly suggest that the scenario is highly unlikely."

MacElroy then proceeded to discuss each of the major concerns of the terraforming project. The killing ultraviolet radiation may still be a problem: "It cannot be determined at this time whether even the most optimistic consequences of an engineering project would decrease the UV flux sufficiently to permit biological survival at the surface. . . ."

The question of available materials has been answered positively: "Access to the minerals necessary for life would seem not to be a problem since, as expected, the elemental composition of the Martian surface is very similar to that of Earth. . . ." MacElroy notes that sufficient water appears to have been indicated (underground since "the polar caps appear not to contain sufficient water to do anything more than dampen the surface"), but nitrogen could be the limiting factor: "While nitrogen may exist beneath the surface, it would remain inaccessible to life, even after a major engineering effort. It is very unlikely that the amount of nitrogen in the atmosphere could ever be sufficiently concentrated by biological processes to levels that would permit life to sustain itself."

MacElroy's conclusion (and his review of the concept was assisted by many members of the old study group, particularly Dr. Melvin Averner) is unquestionably negative: "It is not a strong possibility that even a massive engineering project could trigger the conversion of all Mars into a more hospitable environmental state," even though small self-contained colonies are feasible. "In conclusion, limitations in the concentrations of available nitrogen and water on Mars appear to eliminate the possibility of ecosynthesis on Mars as a whole. These limitations apply directly to biological requirements, and do not affect the feasibility (or lack of it) of engineering the conversion of the present climate to a more temperate equilibrium. . . ."[6]

Is There Enough Nitrogen?

Such pessimism, however, is not universally accepted. The alleged insufficiency of nitrogen, for example, was disputed by Christopher McKay of the University of Colorado's "Mars Project": "Don't forget that the amount of nitrogen in Earth's atmosphere is supported by biological activity, without which practically all of Earth's free nitrogen would wind up locked away in the soil and in the oceans. Assuming as we do now that there is no active biology on Mars, one should have expected that most of Mars' nitrogen would be absent from the atmosphere, and could certainly be in biologically-accessible places not measured by the Viking instruments. Other Viking atmospheric analyses result in models of outgassing that call for a great deal more nitrogen than has been detected, and we know it could not have escaped into space."[7]

Nor, ultimately, should the possibility of importing nitrogen (perhaps from Titan or Triton or some other distant space snowball) be ignored. Sufficient nitrogen could be contained in objects approximately 200 kilometers in diameter, which could reasonably be expected to be movable early in a terraforming program.

Despite what MacElroy said, other Viking data which has relevance to our planetary engineering prospects looks better—except that we've also thought of some other problems previously overlooked.

Energy Budgets

A major ingredient for terraforming Mars is energy. Cleverness and subtlety and superalgae may help, but energy is the bottom line. The amount needed, and the various ways it can be delivered, can help provide an upper limit to the time span needed for terraforming the planet.

To create an Earth-equivalent oxygen layer would require the conversion of about 10 feet of water over the entire globe. The energy needed by the chemical processes breaking the water apart, even if completely efficient, would be equivalent to all of the solar energy reaching Mars for three years. None could be reflected; none used to heat the air or the soil. It all would have to go into the electrolysis of the water. Of course, if you take only 1% of the sunlight, your time scale goes up to 300 years.

The NASA Mars study suggested, based on 1975 knowledge, that the complete volatilization of all the frozen carbon dioxide on Mars could produce an atmosphere thicker than that of Earth, which would create a greenhouse bonus of between 10 and 20°C. The energy needed would be the equivalent of another three Martian years worth of sunlight, totally utilized.

In a more detailed analysis of this limiting factor, based purely on a conservation-of-energy point of view, atmosphere specialist Christopher McKay told the March 1979 terraforming colloquium that the total energy needed was more on the order of *fifty* years worth of *total* Martian insolation. He produced his oxygen from the dissociation of carbon dioxide, a process which absorbed a major part of the whole energy budget. Equally demanding were the processes which released nitrogen in appropriate compounds into the atmosphere. The actual warming of the material was not a particularly large fraction of the total.

Based on a rough estimate for biological efficiency in the use of solar energy, about 0.1%, McKay suggested that 50,000 years was thus a lower limit on the time required, if pure biology was to be relied on.

Other energy sources are possible. The Ehricke proposal to use thermonuclear bombs, McKay mentioned, is "preposterous," since at least 10 million megaton bombs would be needed just to start the process. But solar energy could be increased, particularly through the use of orbiting mirrors (à la Arthur Clarke) to concentrate additional sunlight onto the planet. Together with efforts to increase the energy absorption of the surface, such techniques could reduce the project duration to "less than several thousand years."

More Problems

McKay cautioned that emplacing a one-bar atmosphere on Mars might be only the beginning of the difficult work in terraforming the planet. Since the surface gravity is only one-third of the Earth's, three times as much mass per unit area would be needed. Since the surface area of Mars is a quarter of Earth's, its atmospheric mass would be almost the same.

There are some benefits to this, and some problems. Denser air can provide a more powerful advective heating effect to keep the polar regions warm. But this additional height of the atmosphere (clouds would generally be three times as high) might not allow a stable non-mixing ozone layer to form—and our planned ultraviolet shield consequently might never work. Additionally, the vertically stretched atmosphere would greatly reduce the possibility of orographically induced rainfall, since the ratio of mountain height to cloud height would be less than on Earth.

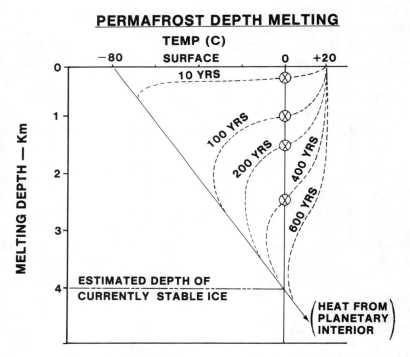

Even if a warm surface climate is brought to Mars, it will be thousands of years before the deep permafrost melts. As it melts it will contribute to ground instability and to the gradual filling of the Martian oceans. This graph by Dr. Edwin L. Strickland III shows the way the melting line will gradually become deeper and deeper.

Although the mass of the atmosphere would not be likely to escape over geologic time periods (say, hundreds of millions of years), there would be conditions in which the air would threaten to freeze out. This trouble, which would not happen for tens of thousands of years after the initial settlement of the planet, is due to the high wobble of the Martian axis, which varies up and down by about 20%. Since the polar conditions control circulation around the entire world, and since planetary climate is determined by the extremes in the polar areas, a highly tilted Mars and extremely cold poles could drive the artificial Earth-like conditions haywire—unless some preventative planetary maintenance were scheduled, such as the "Burns-Harwit Maneuver" described in the last chapter.

Another problem has been described by Dr. Edwin L. Strickland III, a geologist at Washington University in St. Louis,[8] who warns of the stability weaknesses of a Martian surface under which a permanent ice layer is slowly melting deeper and deeper, there is no way to speed it up and it could take millennia to melt completely. There could be constant mudslides, floods from erupting underground water, and other drastic shakeups of the temperate surface layers by the thawing depths.

MacElroy noted this, too: "If ice is buried beneath the surface of Mars, it is well insulated from the heat of the atmosphere and long periods of time would be required at higher temperatures to melt it." There are no shortcuts here!

The "Tidy Coincidence"

McKay pointed out a very curious fact about the effects of a one-bar atmosphere appearing on Mars, which has one-third the surface gravity of Earth. To get the same pressure, three times the mass per unit area is, of course, needed: "The decreased surface gravity of Mars results in [the need for] a thicker atmosphere and an increased greenhouse effect compensating for its increased distance from the Sun. . . . A very tidy coincidence, indeed!" That is, a one-bar atmosphere, once established on Mars, would be thick enough so that no further energy inputs beyond normal sunlight would be needed! It would be stable without artificial support.

"These changes have many meteorological implications for a terraformed Mars," McKay continued. "The Hadley-cell circulation, the level of cloud formation, the height of the tropopause, and the altitude of the ozone layer are some of the phenomena which would be affected.

A complete analysis also needs to include the possibly significant difference in the radius of Mars. . . ." These questions are still to be answered.

Taking Stock

Earlier scenarios for melting the Martian permafrost or for importing volatiles from deep in space remain viable, even after the analysis of Viking data. Details about the state of water on Mars, about the chemistry of the surface soil, and about the concentration of salts in that soil are still to be determined; they are crucial for the planning of a future artificial course for the planet.

Two other important questions deal with the nitrogen cycle on Mars and with the different ways of dealing with ultraviolet radiation. The answers cannot be presented here. What is important is to put those questions in the proper context which I have tried to describe—the context of the ecological cycles of a functioning biosphere, be it on Earth or someday on Mars.

Back to Biology

Assuming we now have the thick air, moderate temperatures, and abundant water, perhaps we can begin seeding the planet in earnest. As realized by the NASA Mars study, a purely biological approach to Mars would take altogether too much time, so we have been investigating physical techniques to create more hospitable conditions under which the biology can have a much better chance to carry out the needed atmospheric transformations, and in a much shorter time.

How long? Again, the NASA study reports: "If the planet were ideal for the growth of algae, and all of the surface could be utilized, with unlimited water and CO_2, a period of less than 100 years might be needed." But the NASA scientists admitted that a more realistic time frame, assuming the selection of an efficient organism, some increase in temperature, a moderate availability of water, CO_2, and nitrogen, and an exponential increase in the oxygen production rate, would seem to be on the order of several thousand years.[9]

Biologist Penelope Boston described various biological tools for terraforming at the Houston colloquium early in 1979. She repeated the importance of the traditional pioneering role played by lichens in the opening up of previously uninhabited areas for biological activities. There are many varieties of lichens, some needing little oxygen. If the terrestrial selection is not wide enough, better species can be engineered in laboratories such as the one in which Boston works.

A few more words on lichens are in order. The lichen organism is in fact a composite of fungal (or *mycobiont*) and algal (or *phycobiont*) cells which live in a symbiotic relationship. The two types of cells growing together develop a body or *thallus,* which is not characteristic of the independent growth of either component.

Lichens are extremely tolerant to lack of water, which is why would-be Mars terraformers regard them with all the affection usually reserved for pet dogs and cats. Several species were kept dry for a year and a half in permanent light, and then a year and a half in permanent darkness, and after eight days of cultivation on agar medium, they produced reproductive spores. However, even the most drought-resistant lichens seem to need a lot of water to reactivate their photosynthesis. Without such sporadic wet periods, the organism would eventually perish.

Boston has also described other candidates for engineering into ideal Mars inhabitants. Mites live happily in Arctic pools with their own built-in antifreeze. There is a red algae which lives on the surface of snow in the Colorado mountains and actually melts the underlying snow to obtain metabolic water.

Recent discoveries of *endolithic* (or "inside rocks") microbial life in the Antarctic dry valleys of Southern Victoria are also encouraging. A primitive form of lichen, it colonizes rock and lives a few millimeters below the crust on north-facing exposures. In this environment, temperature and water conditions are significantly different from the hostile outer climate.

The studies at the Laboratory on Atmospheric and Space Physics concentrate on low pressure biology, and in particular on the process of nitrogen fixation. A special "Mars Project" at the LASP runs a series of test "Mars jars" reproducing known Martian conditions as well as various levels of modified Martian conditions. It has been found that the plants do quite well at high carbon dioxide to oxygen ratios, ratios which would kill animals.

These studies, which Boston believes can lead to low pressure Martian greenhouses providing food supplies for manned Mars expeditions, are still dealing with individual organisms. The next trick is to set up a *system* of organisms. And here the warnings become more urgent. According to Robert MacElroy, co-chairman of the NASA study, "any system that involves interactions between two or more organisms is very complex. On a planetary scale, the management of such an ecosystem will require, first, a major research effort just to describe it, and second, extensive modeling to understand it."

Potential biological factors useful for Mars-loving organisms were outlined by Boston. One function is "anhydrobiosis" (or waterless life), a rather specialized form of dormancy, which is exhibited by many terrestrial organisms already mentioned. "In the anhydrobiotic state," Boston reported, "many plant species can survive to the remarkably low temperature of $-200°C$; in fact, in some organisms, resistance to 0.008 K [degrees above absolute zero] has been recorded." And such organisms can quickly resume metabolism and growth when temperatures climb again.

"A cold-adaptive mechanism even more widely used, particularly in the plant kingdom, is that of *cryoprotection* [italics added], the use of anti-freeze substances to lower the freezing point of cellular fluids. . . . Some organisms deal with low temperatures by shunting cellular fluid to intercellular spaces where freezing will not rupture or damage the cells." All of these techniques could be utilized for the biological pioneers on Mars.

Laboratory work has already begun on screening out candidate organisms, Boston announced, and a much expanded program was justified: "If life catches hold on Mars, the resulting changes in the atmosphere and the total condition of the planet will be of great significance in understanding the interactions of the biosphere and strictly physical processes which occur on Earth. The enormous amount of information that can be gleaned from investigations of this sort applicable to terrestrial biology, and the immense potential benefit to any scientific colonies using some of the Mars-adapted organisms in their life-support efforts render basic research conducted in this area imperative."

An Ocean for Mars?

The Mars project at the University of Colorado at Boulder, while not complete by any means, can be said to have come to fruition in the spring of 1979 when half a dozen students travelled to Houston for the special terraforming colloquium held under the auspices (unofficially) of the Tenth Lunar and Planetary Science Conference. Introducing the program, biology graduate student Penelope Boston (who received her doctorate in 1981) had told the late night audience that "we have tried to take terraforming out of the pages of science fiction and into the realm of science."

Boston continued.

> We have found a recurring flaw in the approaches that we have surveyed which we have tried to guard against in our own efforts. This is the tendency to unconsciously take an Earth-centric view of Mars and to conceive of the ultimate fruits of our planetary engineering as merely a carbon-copy Earth. Mars can never be a facsimile of Earth, but in a significant sense it *can* be made habitable.
>
> For example, the question of oceans on Mars lends itself to this kind of critical analysis. Despite the fact the Earth's surface is three-quarters water [covered], oceans are not an essential feature of biotic systems in general. On Earth, the oceans are essential because the ecosystem has evolved under these specific conditions, and a good case can be made for oceans as a necessity for the spontaneous formation of life. But on Mars we have the option of introducing life at the optimal level of adaptive organization to meet prevailing conditions. The absence of large bodies of water will cause the biosystem to be significantly different from that of Earth, and herein lies the trap.[10]

Boston's disavowal of the need for oceans, and her denigration of their essential role in any modification of terrestrial-type biospheres, led to some dispute both during the session and afterwards. As we have seen in earlier chapters, oceans provide the ultimate source of all rainwater on a planet, as well as a *thermal flywheel* to preclude extreme fluctuations in temperature.

This function of planetary oceans was obvious to Isaac Asimov who, in *Planets for Man* (1964)—a popularization of Rand Corporation scientist Stephen Dole's earlier study, "Habitable Planets for Man"—wrote: "It can be said categorically. . . . that a habitable planet must have fairly large open bodies of liquid water. This is true even for land life, for not only did land life originate in the ocean, but without oceans there could be no extensive precipitation and hence no salt-free ground water to provide the supplies of fresh water upon which all land life depends."[11] Dole repeated that absolute requirement in 1979: "Asimov was rewording what I had already written," he told me, "but the assertion remains correct: without large open bodies of water, you will not get fresh water, which comes only from evaporation and subsequent precipitation."[12]

Recall that there *is* another source of atmospheric water: plant transpiration. On Earth it is not dominant, but on an oceanless planet it could be made to fill some of the gap, perhaps. Also, various species of plant life can indeed survive on highly saline ground water. So a Martian ecology without oceans, à la Boston, *is* conceivable. But recalling facts about Earth's biosphere, such as the deficit of land oxygen supplies being made up for by oxygen produced by phytoplankton in the oceans, it should be clear that a planet without oceans will have land life very different than the land life on Earth. Perhaps some species could tolerate such conditions, but no credible models for an oceanless world have yet been put forward.

Nevertheless, this claim was further defended by Boston: "If we tried and succeeded in creating oceans on Mars, we would probably have doomed the planet in biological terms. On Mars, there are no plate tectonics as far as we can tell. On Earth, the sulfur, phosphorous, and to an extent the calcium cycles are maintained over the space of geological time by a slow recycling through Earth's plate tectonic system. On Mars these biologically precious elements would wash into the oceans becoming more and more scarce on land and finally entirely unavailable to the biosphere. The oceans would become progressively more and more saline. Sediments would wash into the ocean basins with no hope of seeing dry land again and eventually the basins would fill up, becoming very shallow."

Opposition arguments suggested that Boston had taken a much too extended view of the evolution of Mars. One participant (this writer)

Thin layer of snow covers rocks near Viking Lander 2 on the plains of Utopia. When Mars is terraformed, this region will be under half a mile of water, while thicker snow fields will form permanently on high mountains.

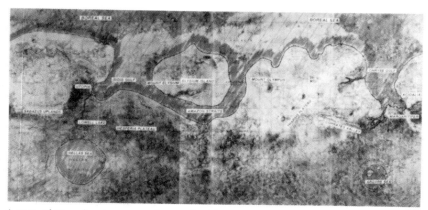

A map of Mars has been superimposed with water at a constant topographic contour, giving an approximate idea of how the Martian oceans might look. Many coastal areas underlain by permafrost could be expected to gradually melt away.

suggested it was wrong to assume that humans would engage in planetary engineering *once*, but then would be unable to set up giant machines capable of maintaining preferred levels of ocean salinity by extracting these precious biological elements. The ultimate disaster Boston predicted might take a billion years to develop, and anything can happen, or be made to happen, in a billion years.

Oceans of Mars

Rains can come again to Mars, and the water can trickle over the land, filling the gullies, dry now for eons. The new creeks, streams, and rivers can run downhill carrying the waters into basins which could form the new oceans.

Topographic maps of Mars can show where the new seas will form and, depending on the amount of water, how far they will extend. Lessons learned the hard way on Earth can teach us how such oceans will play an integral part of the Martian biosphere, and how they will affect planet-wide rainfall, temperature, and soil fertility.

As a case study, let us flood Mars up to the base elevation marked on topographic charts. On NASA charts[13] that line is already marked "0," corresponding to a pressure of 6.1 millibars, but I have chosen it as a hypothetical shoreline for other reasons: first, it allows two arms of the northern circumpolar sea to reach the equator, so ocean currents can help distribute heat; second, it creates some wide shallow seas, giving the most ocean surface area possible per unit volume. Third, although up to 35% of the planet's surface winds up as ocean, the volume of water is still only a tiny fraction of that on Earth—an "economy size" ocean.

Three major bodies of water will form: the *Boreal Sea,* which encircles the northern hemisphere; the *Hellas Sea,* an isolated body of water in the southern hemisphere; and the *Australian Sea,* centered at the south pole. If other isolated basins at higher altitudes also fill up, then more isolated lakes and seas would appear, the largest of which would be the *Argyre Sea.* The southern highlands could be dotted with salt lakes.

In the Boreal Sea will be two small continents, one at middle northern latitudes containing the Elysium volcano and one centered at the north pole (to be called "Arctica"?). Reaching across the equator is an extension of the great northern ocean, the Chryse Gulf, which is bounded on the east and south by extensive lowlands, and which adjoins the flooded Coprates Chasm to the west. Elsewhere the coastal regions generally rise fairly rapidly from the seashore, except on the Utopia Plain, a mid-latitude plain west of Elysium warmed by ocean currents from the Amazon straits. The isolated seas of the south, Hellas and Argyre, can fill to some significant depths before overflowing and spilling north and south. Once such an overflow begins, water erosion could be expected to rapidly cut a deeper channel, allowing a significant amount of formerly isolated water to pour into the Boreal Sea. Such a flow, similar to the so-called Injection Events when Earth's world ocean received influxes of water from previously disconnected ocean basins, could be an ecological disaster on a planet-wide scale because of the sudden changes in surface salinity they would cause.[14]

The water of the Hellas-Argyre Seas would probably be saltier and warmer than the Boreal Sea, and would intermix primarily in the upper layers. When this happened on Earth, mass extinctions and climate upheavals resulted (since the ocean's thermal stabilization mechanism was interfered with). Similar disastrous consequences could be expected on Mars, if two large, long-separated bodies of water suddenly intermixed.

To avoid such consequences, it may be necessary to restrict the rainfall in the southern hemisphere, or to create the long-legendary planet-wide canals so as to allow drainage on a routine rather than sporadic (and catastrophic) basis. The depth of water collection in the Australian, Hellas, and Argyre basins may thus be much less than topographic elevation might suggest.

What about rainfall patterns? Based on the fundamental mechanisms of winds and rain, the following view of Martian climate seems plausible.

The rising equatorial air, replaced by colder air from both north and south, drops its moisture in equatorial rainstorms and then heads for the poles. Considering the shorter equator-pole distance on Mars and the greater vertical scale of the atmosphere, there might be only a single convection cell per hemisphere: that is, the dry poleward-moving air from the equator could make it all the way to the polar regions before descending. The mid-latitude deserts which characterize Earth will then be absent from Mars; instead, the polar regions would contain the deserts, with an ice-free Australian ocean softening the climate around the south pole.

The descending air would pick up moisture and flow back towards the equator, carrying the moisture and being deflected towards the west by the "Coriolis effect." Such moist air masses, encountering rising landforms, would produce steady, gentle rains through the orographic rain mechanism. Regions such as eastern Elysium and the western coast of the Chryse Gulf would be particularly well-watered by these northeasterly wind patterns. The Tharsis plateau might be expected to form glaciers.

In the northern summer, both the Chryse Gulf and the Amazon straits between Elysium and the mainland could be expected to spawn hurricanes which could locally counteract the prevailing northeasterlies. Such storms might be able to circle the entire planet along the pole-girdling Boreal Sea.

Professional climatologists may be able to predict the ocean currents and the seasonal variations of such a newborn planet. They may be able to forecast the extent of glacier-building on Arctica, and the effects of pack ice drifting south through the Boreal Sea towards the temperate zones. They will also concern themselves with transport of windblown dust, with jet streams, and with fog banks caused by clashing currents. Such calculations must lie beyond this present edition, but they will be made.

The essence of the story for would-be Mars rebuilders is that the creation of oceans creates in turn an entire pattern of winds, rain, rivers, currents, and glaciation. The lessons of Earth's complex biosphere must be appreciated and applied, and appropriate computer simulations must map the annual climatic fluctuations in order to determine the optimal ecology.

194   New Earths

A Scenario for Mars

It's easy to see what's wrong with Mars today, as far as a habitable Earth-like environment is concerned. The Viking Project has given us plenty of data with which we can plot the past and present climate of the planet and our imaginations can give us visions with which to contemplate the future climate of Mars.

People have not yet even reached Mars. Depending on political climates (it is not a question of technology), the first human footstep could be made on Mars well before the end of the millennium. These explorers could extract their oxygen and water locally and set up greenhouses to grow much of their own food; these greenhouses could include the first fractional acres of a future terraformed Mars.

Decades or centuries later, a series of events could occur which would measure the extent of human possession of Mars. As the population grows, children will be born. As the air thickens, spacesuits will rust into disuse in closets. As the waters return, the first rainbow in billions of years will be seen by native-born Martian men and women.

In the edition of Genesis which applies to Earth, the rainbow was

A rainbow appears over the Vallis Marineris, as thunderhead clouds form in the distance. Rains will come again to Mars, and the rainbow will be the symbol of the return of these waters.

the symbol of a promise to withhold worldwide floods and to restrain the waters. In the edition of Genesis for Mars, the rainbow will mean just the opposite: the waters are now free to spread across the planet and prepare it for the dawn of life.

> *And if the sky and sea be foes, we will tame the sea and sky.*
>
> The Ballad of the White Horse
> G. K. Chesterton

9

Venus as a True Twin

The one thing that really annoyed him was the absence of surf. **The skies are pretty good now, and I can accept the barren shores because even green Earth has its barren stretches,** *he thought slowly.* **But this is a hell of an excuse for an ocean!**

Cruising at an altitude of about 500 feet, the man followed the shoreline ahead of him into the misty middistance. The water's surface was nearly motionless, which might have seemed a contradiction since definite traces of water erosion stretched for more than a mile inland. But the pilot knew the reason.

Feast or famine, he snorted to himself. Either the water lay comatose in its basin, or it was driven into an insane fury by the shadow storms that accompanied the manipulations of the giant space parasols a thousand miles above the planet. In neither case did it look much like the facsimile of Earth that was promised to appear.

A glance at his outside instruments assured the pilot that terrestrial conditions were still a figment of the long-range planners' imaginations. The temperature read 312°F, with an air pressure of 15 bars—fifteen times the pressure back on Earth. It didn't show on his instruments, but the pilot knew the outside air was still totally unbreathable: 20% nitrogen, wisps of argon, and all the rest carbon dioxide. The only free oxygen within 200 miles was in his cabin.

The pilot considered a graph somebody had once scratched into the wall above the instrument panel. It showed the temperatures dropping year by year, perhaps a total of 100 degrees in the thirty years people had been established in the atmospheric aerostats; a similar line showed pressure dropping by more than 15 bars.

By the time I die of old age I'll be able to boil my coffee by sticking it outside the door, he thought bitterly. *The conditions will then be just barely fatal, instead of overwhelmingly so.*

His target for today's reconnaissance appeared ahead of him: a large river delta criss-crossed by small streams and littered with boulders carried downstream by last year's floods. That was what the water was needed for, even if it was still far too hot for terrestrial organisms. It had billions of years worth of erosion to accomplish in less than a century, in order to make the planet useful.

The scout ship, which was more of a dirigible than a winged aircraft, turned inland to trace the upstream course of the river. Instruments high above the atmosphere were still unable to pierce the gloom and measure the things which the planet rebuilders needed to know: river flow rates, temperatures, silt composition, dissolved gases in the water. Former attempts to leave permanent instruments had resulted only in piles of scrap metal in the bottom of the ocean—the sporadic storms were still too violent.

Off to both sides of the river, the pilot could see forests of geysers spouting several hundred feet into the air. *The water never gives up,* he realized, considering the constant clash of river and still-glowing surface rocks only a few hundred feet underground. But each new burst of steam lowered the temperature of a few ounces of rock a few fractions of a degree.

He tried to envision the valley as it would be a hundred years in the future: green with plants, perhaps sectioned off into farms, perhaps left wild for grazing or for recreation. He had stopped being skeptical about the project's success, stopped even wishing for some sudden disaster to allow old Venus to reassert herself and shrug off the unaccus-

tomed trappings with which alien visitors were draping her. I guess it will happen just about like they say, *he shrugged.* But I'll be double-damned if they will make me happy about it.

When they lie back on the grass that will someday grow here, I hope they can look back a century and imagine me drifting by. *They would never, of course, be able to imagine him as an individual personality, but some records would survive, some remembrance.*

The future inhabitants of terraformed Venus would have other concerns to bother them. They would spare little sympathy for him, for his crimes he still had not repented of, for his life sentence in lieu of execution, for the surgically emplanted blood cooler still humming in his chest, for the brain stimulator circuits which assured his total obedience short of death.

Damn them and damn their planet, *he suddenly burst out.* Damn their new planet, I hope they choke on it, *he cursed again, reliving the moment in his life which had condemned him to this hell world. The thought that his bones would contribute to the crops which would feed the settlers was intolerable to him, but there was no way out. All the doors were locked; they held all the keys.*

Venus, 2351 A.D.

As far as planets go, perhaps the old proverb is correct and "beauty is only skin deep." At least, that is certainly true for Venus, the brilliant jewel of our dawn and dusk skies. Beneath that superficial beauty lies a world with conditions as close to those of Hell as we may ever find in the universe.

The air seems denser than the water in Earth's deepest oceans. It consists of superheated carbon dioxide at a pressure nearly a hundred times that of Earth's atmosphere, flavored with sulphuric acid stronger than that from an automobile battery. There are traces of nitrogen, argon, and water vapor which help contribute to the most effective "heat trap" in the Solar System. The slow Venusian "winds" are more like ocean currents in their gentle inexorability.

The prospect of what would happen to an unprotected human body on Venus is ghastly to imagine. No breath would enter the crushed chest as the ribs cracked and the diaphragm ruptured. Against the inward crunch of the carbon dioxide sea, the explosive force of boiling body

Venus as a True Twin   199

The hellish conditions on today's Venus would mean instant death for any Earth life form. Even specially designed robot probes break down within hours.

fluids would shave off flesh, layer by layer. These body fragments would char and sear in the heat before drifting off on the currents. Within moments, all that would be left would be calcium deposits and smears of soot atop the glowing rocks.

As late as thirty years ago, such conditions were still unknown. A prominent astronomer remarked in the late 1940's, concerning the surface conditions of Venus: "It may be like the Bahamas, or it could be sudden death if we stood on its surface."[84]

Indeed, today we know it would be sudden death. But it may not always be so.

Venus has been called the "twin" of Earth, due mainly to its size and density. For centuries, people wondered about conditions below those clouds, and wondered if such an idea as terraforming would even be necessary.

Since Venus has an albedo of 0.71, astronomers were able to calculate that the amount of solar energy actually reaching the planet's surface was about the same as on Earth. If the atmospheric constituents were similar to Earth's, then the surface conditions might also be similar. (It now seems obvious that had that assumption been true, the albedo of the planet would not have been so high—but hindsight is always better.)

Many different visions of the Venusian surface were conjured up. Besides the "Bahamas" image (or its variant, the "dinosaurian swamps" image), conceptions of planet-wide oceans, or planet-wide deserts, or of planet-wide hydrocarbon "oil spills" were also debated.

In fact, the first spacecraft designed to reach the surface of Venus were built by engineers expecting to encounter a sea of petroleum. The Soviet Union's *Venera* probes of the mid-1960's were equipped with plugs of sugar which would be dissolved by the petroleum, releasing springloaded antennas.

Needless to day, it never worked. The probes were fried and crushed by the unexpectedly hostile Venusian conditions long before they reached the surface. After several years of failures, and after the results of successful American fly-by probes and ground-based readings, a more realistic picture of the planet emerged.

The "air" on Venus (perhaps it would be more appropriate to call the suffocating blanket of superhot, dense carbon dioxide an "ocean") is ninety-one times as thick as on Earth, and it's almost entirely made up of carbon dioxide. As measured by the Pioneer-Venus 2 fleet in December, 1978, its composition is as follows: carbon dioxide, 97%; nitrogen, 1–3%; helium, .025%; neon, less than .025%; argon, less than .020%. Below the clouds, water makes up .1–.4% of the air, along with .024% sulfur dioxide and .006% oxygen. That's enough water for a layer 10 meters thick planet-wide.

The abundances of nitrogen and carbon dioxide seem to be similar to those of the Earth; water, however, is conspicuously in small supply. If Venus ever had large amounts of water (many planetary geologists believe there is good reason to suspect that it did), the water is gone, perhaps dissociated by sunlight into oxygen and hydrogen. The oxygen would be locked up in the surface rocks, while the hydrogen would escape into space over several billion years. Pioneer-Venus found that the hydrogen-loss rate today is very, very low: water loss from Venus has essentially ceased.

The Shape of Venus

The physical properties of Venus have been easy to determine. It is slightly smaller than Earth (6100 kilometers in radius rather than 6400), slightly less dense (5.27 versus 5.52—corrected for gravitational com-

Venus as a True Twin

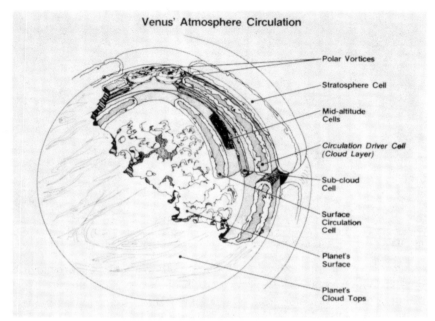

The choking atmosphere on Venus holds in the heat via a "runaway greenhouse effect."

paction, the figures are 4.40 and 4.45). Its distance from the Sun averages 0.7 times Earth's distance (so it gets 100% more sunlight hitting its atmosphere). Its surface gravity is 90% of Earth's.

Radar studies from Earth revealed that Venus hardly rotates at all. It has a slight spin in the opposite direction of the other planets, making one complete turn every 224 days; combined with its motion along its orbit, this gives a sunrise-to-sunrise "day" of 118 Earth days. Oddly enough, this rotation rate is just right so that every time Venus comes closest to Earth (astronomers call it *inferior conjunction*), the same side is facing Earth; this occurs once every 684 days (astronomers call this period an Earth-Venus synodic year).

Explanations for the anomalous rotation of Venus and its bizarre "lock-on" to Earth have been typified by desperation. Perhaps Venus was hit by a giant asteroid. But why then is its axis still nearly perpendicular to its orbit? Perhaps over the eons, the sun-driven winds literally blew the planet's surface backwards. Perhaps the "lock-on" resulted from formerly closer orbits of Earth and Venus. But that would not work because they would have had different orbital periods, and the present lock-on

Without clouds, Venus would appear as a relatively smooth planet, without the towering mountain ranges and deep ocean basins created on Earth by continental drift.

is a result of the present values for their orbital periods. Perhaps the material which condensed in this portion of the Solar System was too turbulent to give a preferred angular momentum to the resulting planet. But the same objection to the first guess also applies here.

A few years ago, I suggested (tongue-in-cheek) that the lock-on was deliberate, the result of efforts by the former inhabitants of Venus who, in a desperate interplanetary war with Earth fought by laser, altered the spin of their planet so that at closest range at least half of their world was safe from Earth's death rays. Since Earth is still habitable and Venus is not, I facetiously theorized that we must have won the war.

The suggestion (in *Astronomy* magazine) was greeted with all the attention it deserved—absolutely none[1]. But I raise it again now to illustrate my intuitive belief that whatever the explanation for this anomalous rotation proves to be—no matter how remote from our current scientific imaginings—the implications for future planetary engineering feats will be substantial. But time will tell.

Our knowledge of the geology of Venus is still based primarily on radar observations from Earth and from the orbit above Venus, augmented by photographs and radiation readings by Soviet robot landers. Together they have revealed some interesting details about the shrouded planet which have enabled geologists to make some generalizations, leading to theories of the planet's formation and evolution.

Venus is a relatively smooth planet (similar to the smoothness of lunar maria) with local areas of significant roughness and greater relief.[2]

Venus as a True Twin   203

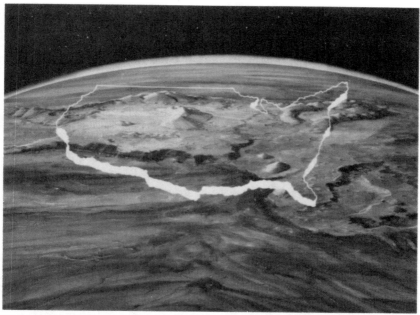

Two America-sized continents have been charted by radar pulses from an orbiting spacecraft. As with other geographic place names on Venus, they have been named after females: Ishtar and Aphrodite.

The most abundant features visible to radar are what appear to be large craters, similar in shape to large degraded impact craters on Mars and the Moon. The second most abundant features are isolated peaks and mountains, found in nearly every area imaged by radar (craters are not universally distributed—some regions seem to have destroyed them if they existed), and probably created by volcanism. Some mountain ranges have been detected, implying at least minimal continental drift; a large trough-like depression 1400 kilometers long by 150 kilometers wide by 2 kilometers deep, running across the equator trending approximately north-south, reminds geologists of Earth's Afro-Arabian rift valley and the Martian Valles Marineris, possibly further evidence of other tectonic activity. And one highly radar reflective structure, named "Beta," seems to be a giant volcanic caldera 10 kilometers high, 700 kilometers in diameter, with a 60-by-90-kilometer depression at the summit.

The Russian probes which landed in 1975 added some interesting data. Venera 9 landed in a region with an abundance of slab-like and rounded, dull grey, massive boulders, typically half a meter across, resting on fine-grained dark soil. The Venera 10 site showed large flat platforms of outcrop separated by large connected areas of soil; some fine dark material has been trapped in pits on the rocks. Analysis of data from the one Pioneer-Venus probe, which survived on the surface, also points to the existence of a fine dusty soil which was splashed into the air by the probe's impact and which took three minutes to settle.

A History of Venusian Climate

The history of Venus' climate is largely a matter of conjecture, since high-resolution maps of the surface (which could show effects of running water or continental drift or glaciers) are yet to become available, and chemical analysis of the present atmosphere is in its infancy. Some scientists suspect that Venus was created in nearly its present form and climate; others suspect that conditions a few billion years ago may have been far more comfortable, even to the point of allowing surface water to gather, and perhaps even allowing the development of living creatures.

The transition from a formerly habitable (perhaps inhabited) planet to a sterile oven would have taken place catastrophically—not over a gradual process spanning millions of years, but, on planetary scales, practically immediately (within a year, or perhaps within a few days). Once a planet's biosphere begins to slip over the "knee" of the curve, disaster is instantaneous. The notion that such a fate could befall Earth, unless the trend were recognized and headed off by planetary engineering, is terrifying, and a cogent plea for the development of terraforming capabilities.

According to Dr. James Pollack, any guesses about the early climate of Venus depend strongly on what kind of atmosphere the planet possessed.[3] Taking into account the 30% lower solar brightness, computations show that a reasonable atmosphere on Venus would have led to "quite moderate surface temperatures" for more than a billion years; furthermore, Pollack suggests that "runaway conditions were not reached on Venus until halfway through its history."

Even with great amounts of carbon dioxide, Venus could have had a moderate climate if water was in short supply, but present in enough amounts to form pools within which carbonate rocks could be formed to absorb some carbon dioxide. The addition of off-planet substances in trace amounts (Pollack suggests "collisions with volatile-rich comets and asteroids") may also have had catastrophic climatic effects by blocking off hitherto open "windows" in the "greenhouse effect." All these speculations merely reaffirm one of the first principles of planetary engineering: major climatic effects can be produced by relatively insignificant alterations in the atmospheric composition or energy budget of a planet.

Pollack concludes with an exciting speculation for future reconnaissance of Venus. "If Venus had a moderate surface temperature in its early history and if at least some water was present, then fluvial water-eroded features may have formed at that time and might still be recognizable at present. It will be of extreme interest to examine high resolution radar images of the surface of Venus, as may be obtained from an orbiting spacecraft, to see whether such features are in fact present."

And where there may have been water for a billion years or more, there may have been life. Such a concept of the history of Venus may give an almost mystic justification for terraforming the planet. Far from interfering with the "natural order" of the universe, humanity would be *repairing* a climatic catastrophe which had destroyed a habitable planet. Terraforming Venus to make it Earth-like would then be seen as merely *returning* it to its natural state.

Early Plans to Rebuild Venus

Because Venus already has a thick atmosphere, it has been the subject of much speculation regarding the possibility of engineering it into a more truly twin planet. Since the air is mostly carbon dioxide (beloved gas of biological activities), and since it seems that the carbon dioxide has locked in most of the excessive surface heat, an obvious approach would be to use plants to break down the carbon dioxide and the "greenhouse effect" in one blow.

Such was the plan described in Poul Anderson's science fiction story, The Big Rain (1954). This idea assumed that there was a large supply of water available which, with the dissolution of the "heat trap," would help cool off the surface.

Carl Sagan popularized that notion in his own articles, books, and speeches. Writing in Planets (1966), Sagan summarized how it might happen.[4]

> Dead as it seems today, Venus need not be written off forever as an abode of life. . . . Although Venus may be unable to develop life of its own, it offers a possible field for colonization from Earth. This might be attempted after an intermediate period of what can be called 'microbiological planetary engineering.'
>
> No one who has studied the swarming, varied life of Earth fails to be impressed with its amazing ability to adapt itself to a wide range of environments. Microorganisms thrive in hot springs, on snowfields, in saturated brines, in the pressure, dark, and cold of ocean bottoms. It is not inconceivable that an organism can be found or developed that will live and thrive somewhere on Venus and, in time, make it habitable for higher forms of life.

Sagan goes on to suggest that the clouds of Venus be seeded with these microorganisms where "they must live a wholly aerial life, suspended by the turbulence that stirs the clouds." The most likely choice for such an organism, Sagan continues, is the so-called blue-green algae.

> According to one speculation, [continues Sagan, evidently referring to Anderson's story without naming it] the arrival of micro-Pilgrims in their space-borne 'Mayflower' may have a dramatic effect on Venus. Once they are established and begin to multiply rapidly, they will abstract large quantities of water and carbon dioxide from their cloud homes and they will break down these substances by photosynthesis, manufacturing food and discharging oxygen into the atmosphere. . . .
>
> If these reactions continue long enough—and this need not be very long—they will take much of the carbon dioxide out of the atmosphere, replacing it with oxygen. . . . The surface will then cool down. If it gets cool enough for rain to collect at the poles as liquid water, the greenhouse effect of water vapor in the atmosphere will

The first colonists from Earth may be blue-green algae, shown here being dispensed by a fleet of buoyant balloons high in the Venusian atmosphere. They could break the carbon dioxide down, releasing oxygen.

be reduced also, causing additional cooling. When more rain falls, the heat-retaining clouds will partially clear away, leaving a planet with an oxygen-rich atmosphere and a temperature cool enough to sustain hardy plants and animals from the Earth. It may even be suitable for human colonization."

These tricks with algae sound so tempting that many space enthusiasts seem to consider them only a matter of building a few rockets and of launching the small cannisters of biological terraforming tools to Venus.

Writing in *Galaxy* magazine in 1976, Jerry Pournelle typified such unbridled optimism. "Not only can we terraform Venus, but we could probably get the job done in this century, using present-day technology." The cost, Pournelle guessed, is "unlikely to be greater than a medium-sized war," or about 500 billion dollars. All it takes is the algae, he went on to assure his readers. "In no more than twenty years from Go, the Big Rain will strike the ground. . . . At that point the surface will be tolerable to humans with protective equipment."[5]

This was a glorious vision, based on educated guesses and a bit of legerdemain. A more realistic view, possible in light of our recently acquired data, would not fault the suggestion that organisms could be salted in the Venusian atmosphere, or even that they could thrive there; but it would contradict the cheerful image of breaking clouds and Earth-like climes all due to one biological trick, in a short period of time. More detailed problems with this scenario will be discussed shortly, but one immediate objection deals with the image of clouds breaking to let out the heat: the solid cloud mass reflects most of the Sun's heat, and a smaller cloud cover might have just the opposite effect desired by allowing more heat to enter the atmosphere.

Then, too, the initial efforts might be a little more subtle, seeking that "ecological fulcrum." At the terraforming colloquium in Houston in March 1979, planetologist Steven Welch suggested that the sulfuric acid, not the carbon dioxide, be the first target of terraforming.

"Sulfuric acid enhances the greenhouse effect by plugging radiation holes left by the carbon dioxide and water vapor," he told the colloquium. "Its effect is all out of proportion to its mass—so a modest effort to unplug these holes, using biological tools to remove the sulfuric acid, could have major results in letting heat leak away from the planet."[6] Welch's ideas were very well received, as they marked a major break with the wishful thinking of the past. They opened a new chapter in the planning to rebuild Venus.

What's Wrong With Venus Today?

What are the real prospects for terraforming Venus? To appreciate these problems, let us take a look at the planet from the point of view of Earth life forms. The problems are primarily ones of heat, air mass, and spin.

First and most obvious is the problem of heat. We need temperatures below 40°C, not above 460°C as now exists. This is only partially a function of being closer to the Sun, since present-day Venus reflects so much more sunlight than does Earth that the amount of solar energy reaching the surface of both planets is about the same. However, on Venus that heat is trapped by the "greenhouse effect." In addition, it is not just the air that is hot. The surface rocks are hot, too, to great depths where the heat of the planet's interior begins to dominate.

This leads to the second problem—the too massive atmosphere. Presently consisting mostly of carbon dioxide, this air would be a problem even if it was all converted overnight by magic into oxygen and solid carbon soot. In that case, we would still have three hundred times too much oxygen.

Venus as a True Twin 209

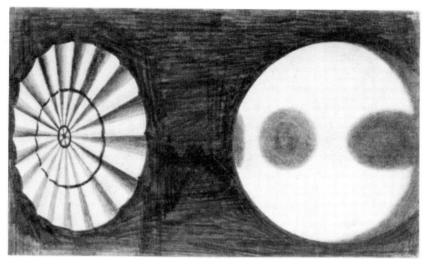

One of the first tasks in terraforming Venus is to block out most of the incoming sunlight for several centuries, perhaps using giant parasols such as these shown here.

Thirdly, the rotation rate of Venus is extremely uncomfortable: sunrise-to-sunrise is about 120 days. This is an awfully long day-night cycle for plants to adapt to, unless (as was suggested at the 1979 Houston terraforming colloquium by Welch) the cycle is considered as a short year rather than a long day. After all, the ecology of arctic and antarctic regions on Earth does tolerate long periods of full sunlight followed by long periods of complete darkness.

An early suggestion to solve the spin problem by impacting space icebergs is no longer tenable (even though it resulted in some exceptionally interesting paintings showing a rapidly spinning Venus with an equatorial ocean and habitable continents clustered around the poles). Calculations show that the momentum involved is not really high enough to significantly alter the spin rate of Venus. And the heat generated by these impacts would be an additional amount of excess energy which would eventually have to be dissipated.

Now suppose that the incoming sunlight could be entirely blocked off for a long period of time. Even then, it could take hundreds of years for the planet to cool off; the limiting case would be if the planet's bare rock radiated its heat directly into space without any interference from an atmosphere. Calculations show that it still would take a few centuries for the surface to drop to survivable levels; an atmosphere with even a small "greenhouse effect" could multiply that time requirement many fold.

The Surplus Air Problem

Imagine that all the carbon dioxide could be converted in the wink of an eye into free oxygen and carbon soot. Now instead of 90 bars of carbon dioxide we have "only" 60 bars of free oxygen, still three hundred times as much as on Earth. To get rid of it, we may try to lock it up in surface rocks (the hotter the rocks, the faster the reaction); but calculations show that the mass of rock needed to absorb that excess mass of oxygen (assuming even that the rock were not already partially or even predominantly oxydized) would cover the planet to a depth of 100 kilometers or more.

That means that the top 100 kilometers of the surface of the entire planet would have to be *gardened,* or selectively exposed to air so as to absorb the oxygen. On Earth, this process is accomplished by water erosion of mountains raised by tectonism over eons of geologic time. Something different would be needed on Venus.

Imagine a scheme to pump the excess atmosphere of Venus off the surface and out into space. Repeated flights of tanker spaceships shuttling between the upper atmosphere where they scoop up gas, liquify it, and store it under tremendous pressures in a special hull, and orbit where they eject the gas at escape velocity are possible, but a more attractive (and more energy efficient) system involves a sky hook.

Our Venusian skyhook operates on the same principles as described previously. A line dangles into the atmosphere of Venus and collects air via a ram scoop. The gases can not, of course, be sucked up the line since vacuum pressure is not nearly powerful enough to raise them hundreds of kilometers to the mother ship. Instead, the gas is tanked (or alternately frozen) and shipped up a "freight elevator" device. At the main terminus satellite, the gas is used as reaction mass to overcome the drag effects of the thin upper Venusian atmosphere, through which the main vehicle is orbiting.

Such a machine would probably not need a cable longer than about 100 kilometers, as compared to geosynchronous skyhooks around Earth, which would need 40,000 kilometers of cable length. And the machine would need a powerful source of energy, but Venus provides one: its proximity to the Sun provides solar power with twice the intensity of near-Earth levels. Given electricity, the terminus satellite could expel the Venusian gases using either a mass driver or a slinger system; either one, by the twenty-fifth century, may be able to provide the almost 20 kilo-

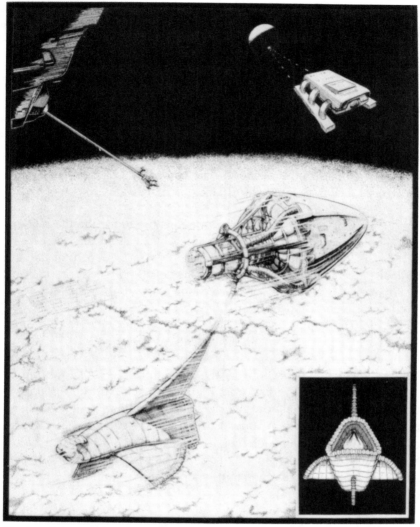

The Venusian atmosphere is much too dense, no matter what its composition; this two-mile-long machine was dreamed up to export large quantities of the excess gas and disperse it into space.

meters per second needed to achieve a Venusian escape trajectory from a low orbit over the planet.

The orbits of the fleet of gas-extractors would probably be in full sunlight, circling over both poles in turn; for Earth, we call this a "sunsynchronous" satellite. Its full sunlight is maintained by orbital torquing caused by the oblateness of Earth; over spherical Venus, some additional course corrections would be periodically needed, provided by the main impulse engines.

The number, capacity, and efficiency of such machines can only be guessed at today. But let's specify an upper limit, just to be conservative.

If we wish to remove 98% of the mass of the Venusian atmosphere in a reasonable time, say 100 years, we must haul up a mass 10 quintillion tons, or 300,000 tons per second. Compare that to the flow rate along the Amazon river, Earth's mightiest watercourse: 10,000 tons per second. The largest machines built which handle flowing water, like the Niagara Power Project, handle 400 tons per second.

Or look at it from an energy requirement: hauling the mass of gas 100 kilometers high, and then accelerating it by 20 kilometers per second, requires about 10^{25} ergs of energy over a 100-year period. That's all the sunlight falling in the same period on an area of 10,000 square kilometers, assuming 100% efficiency of conversion. Throw in a factor of ten for engineering reality, and the fleet of air scoopers must have solar collection panels of 20 million square kilometers, which is three times the total area of Venus.

Let the engineers of the next century concern themselves with the building of such super-scoopers! The strength of the cable and piping system, 100 kilometers long, from the main spaceship to the scoop, is far beyond any material now available. The scoop itself will be gouging through the atmosphere at about Mach 30, the re-entry speed of spaceships. On Earth, such spaceships have heat shields which can barely sustain such temperatures for minutes; on Venus, some mechanism must allow the leading edges of the ram scoop to survive for months or years between maintenance.

There is another way to rid ourselves of this surplus oxygen without removing it from Venus. Scooper machines would also be needed. Recall that our problem was to remove the oxygen from the atmosphere, either up into space or down into the crust. Yet there is not enough material on Venus to absorb even a small fraction of all the liberated oxygen except back into the deadly carbon dioxide—unless hydrogen is imported.

That is the suggestion Pat Rawlings at NASA's Johnson Space Center in Houston came up with while discussing the design of the Venus scooper

Venus as a True Twin 213

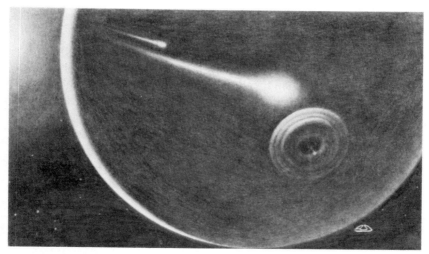

Another way to dispose of the superabundant oxygen would be to import hydrogen from Saturn, which would create water as soon as it hit the Venusian atmosphere.

machine. (A Florida space enthusiast named Burke Carley developed the same strategy independently.) Build the air-grabber machine, yes, but use it to steal hydrogen from Saturn, crate it, and ship it off to Venus where on entry into the atmosphere it will burst into flame, absorb the oxygen, and produce water!

You gain the following: the hydrogen in water weighs only one-eighth as much as the oxygen, so to tie up a given amount of oxygen, only one-eighth of that mass has to be lifted away from a giant planet. Saturn is a candidate because its atmosphere contains hydrogen and helium and because its gravitational attraction leads to an escape velocity comparable to that of Venus.

The hydrogen should be spiced with ammonia from Saturn, or more likely, Titan, which would react with the free Venusian oxygen to yield nitrogen and more water.

Our hypothetical Saturn mining machine, then, has the same kind of scoop on the same long cable dangling from the mother ship circling the planet. Some of the stolen gas (perhaps the fractionated helium) can be used as reaction mass as long as careful navigation shows that the expelled stream of material would not interfere with the stability of the rings of Saturn. The hydrogen can be extracted and removed from the scooper machine by periodic visits to a tanker ship, whose job it would be to transport the hydrogen to Venus.

The most compact form of hydrogen is *metallic hydrogen,* a theoretical form of the gas believed to compose the core of Jupiter and Saturn. Crushed under enormous pressure, the hydrogen molecules (since hydrogen normally exists as the molecule H_2) are stripped of their electrons, broken apart, and crushed together. The resulting material is a powerful conductor of electricity, due to all the free electrons; this leads to the intense magnetic field of Jupiter.

Such a form of hydrogen might be the handiest way to store the gas in transit. An electromagnetic cannon orbiting Saturn could fling tanks of the material towards Jupiter for the tricky "reverse swingby" maneuver in towards Venus. The tank structure, perhaps 1 or 2 kilometers in diameter, would likely be expendable, tearing apart upon entry into the Venusian atmosphere five years after leaving Saturn.

When Can Oceans Form?

All this talk of terribly high temperatures on Venus may seem to preclude any ideas about forming bodies of open water on the surface, but that would be overlooking the second important number which tells at what point water will boil: air pressure. As kitchen pressure-cookers have taught us, water can be kept liquid at temperatures considerably higher than the "sea level" (that is, 1 bar) boiling point if enough pressure is applied.

And there's plenty of air pressure on Venus! In fact, even the present situation (90 bars, 450°C or 860°F) is only 190° above the boiling point of water (260°C or 500°F) at that pressure. That means, if Venus were shielded from enough sunlight to drop the surface temperature just 190°C (30%), and if there were water vapor in the atmosphere, it would begin to rain. The heat of the surface would flash the water into steam, but such a process could be quite an efficient way of cooling down the surface once incoming sunlight had been reduced.

Even after our hypothetical algae had reduced the carbon dioxide to pure oxygen (about 60 bars worth), the boiling point would still be quite high, about 200°C (400°F). Considering the numerous holes punched in the "greenhouse effect" by atmospheric engineering up to this point, it would not be out of the question to expect the surface to be "cool" enough to support large bodies of water. This is the point, recall, at which hydrogen stolen from the outer planets' atmospheres is to begin arriving. Combined with all but a fraction of the 60 bars of

Liquid water may be able to form on the sterile surface of a half-terraformed Venus, even when temperatures are far too high for life, since the air pressure will be very high. The water will erode the surface and prepare the soil for life.

oxygen, it would create enough water to cover the entire planet to a depth of 600 meters (2,000 feet). That's sufficient for some very respectable oceans.

Such oceans of water at 200°C would, of course, be lifeless. Limitations on terrestrial biology are functions not of the boiling point of water but of the point at which organic molecules break down. True, some simple forms of life can survive at 100°C, and genetic manipulations *might* be able able to create precursor micro-organisms which could perform valuable biological processing on the surface once temperatures fell below 150°C. But that's a long way down the road.

So why the hurry to form still-sterile oceans? There is a good reason for such a plan, and it is related to the role of flowing water in the modification of Earth's own surface. Water prepares soils by laying down deposits and dissolving harmful salts. Such activities are driven by rain, which of course comes from evaporation from large open bodies of water. Thus, the sooner oceans and rainfall can form, the sooner the mechanical gardening of the surface dirt can begin, and the closer we will be to creating genuine "soil" from alien "regolith."

Even without life, the dry land need not be motionless. Machines designed to operate under such temperatures and pressures could carry out rock-crushing, river-damming, and canal-digging operations as needed; they could process soil, river water, and air to extract desired materials. Eventually, even before the air is breathable, the machines could plant the new forests on the ancient mountains.

Life will not grab hold of the entire surface at once, of course. Conditions will become endurable first on mountain tops near the poles (which will be the worst place to look for good soil and for plentiful sunlight for photosynthesis), and only gradually will green growing things (tended thus far by unbreathing robots) edge downward into the hot, dense air of the lowlands. Perhaps the newly discovered Maxwell Mountain complex near the north pole will be one of the sites for the first surface vegetation of the planet.

A Strategy for Venus

We have examined possible ways to overcome the thick air and high temperatures of Venus, with the goal of creating a breathable atmosphere and a tolerable climate. Early proposals to terraform Venus by introducing blue-green algae in order to break down the carbon dioxide are now seen to have been concerned with the wrong targets while missing far more important possibilities. Even if these algae proposals had been completely successful, they would not have measurably affected the unhealthy characteristics of the planet.

The most attractive targets in the stifling Venusian "greenhouse effect" are the *trace* components which block important "windows" for the escape of heat (since carbon dioxide *by itself* does not produce a particularly powerful "greenhouse effect"). These are sulfur dioxide and water vapor; both can be attacked either biologically or chemically.

While more heat is being let out of Venus, it would be useful to reduce the amount of incoming heat. This could be accomplished by shading techniques. Since the air and surface rocks of Venus are already so hot, even a cutoff of all sunlight entirely would not cause the temperature to drop significantly for decades. The time scale for the surface rocks to cool to habitable levels, even under the best conditions, would be measured in centuries.

When the time comes to convert the atmospheric carbon dioxide into carbon soot and free oxygen, then the free-floating blue-green algae

The polar uplands of Venus would probably be the first regions to become capable of supporting terrestrial life. Here, a pine forest grows on the slopes of Mount Maxwell near the north pole.

will indeed be useful. But once the free oxygen is released, the problem is only half-solved since there still would be several hundred times too much oxygen for a habitable world.

Some of that oxygen may be absorbed by surface rocks, but without substantial erosion, the rocks will not be able to handle more than a tiny fraction of this oxygen. A good way to eliminate this oxygen would be to import vast quantities of hydrogen from the atmospheres of the outer planets, a tactic which would convert the unwanted oxygen into badly needed water.

Because of the high pressures of the Venusian atmosphere, the boiling point of water is substantially elevated, so oceans of superheated water could form soon after the temperature began to fall. The water would be sterile for centuries, but it would perform other valuable functions needed to prepare Venus for surface biology: it would support rainfall, continental runoff and soil deposition, and other forms of erosion. By the time the surface becomes cool enough for biology, a major part of soil preparation will have already been accomplished by the action of these new Venusian oceans.

Ultimately, the oxygen level will be stabilized somewhere below ½ bar, and nitrogen, somewhere around 1 or 2 bars. Temperatures will drop below 50°C and life will be introduced, first on the high polar mountains, and eventually on the wide lowlands.

The first colonists for Venus can occupy the upper reaches of the atmosphere, living in giant floating aerostats from which they can visit the surface in refrigerated submarine-like ships. At the altitudes at which the aerostats will float, temperature control would not be a problem; in fact, the gondolas would need pressurization and heating. The high altitude winds would sweep the aerostats all the way around the planet every few weeks, although aerostats over the poles would probably maintain their positions without active propulsion.

These aerostats (and there would be hundreds of them) would be the staging areas for the assault on the surface. As the atmosphere evolved, they would drop lower and lower. Eventually, these craft would run aground, ark-like, on the "Mount Ararat" of a new world. Of course, by that time the seeding of the surface would already be well-advanced. The Venusian ocean-atmosphere would be conquered.

Venus, at last a twin of Earth, will become a new home for humanity. Its skies and seas would have been tamed.

Loose Ends and Open Questions

The preceding scenario ignores some substantial roadblocks to making Venus look like a clone of Earth. First of all, the spin of Venus is still too slow for comfort. Secondly, the lack of a natural magnetic field may have detrimental effects on biological activities. Thirdly, the incoming solar radiation will have to be perpetually reduced to maintain a habitable climate so a lot of the sunshades will have to remain in place forever.

On Earth, most of our original carbon dioxide has been locked safely away in carbonate rocks through a process called the *Urey Reaction*. This process works most efficiently under water but is fairly slow. The oceans of Venus will be able to hold a large amount of dissolved carbon dioxide, but we probably cannot use the Urey Reaction to permanently remove substantial quantities of the unwanted carbon dioxide.

The need to keep sunshades in place may suggest a way to create a 24-hour day-night cycle *without* changing the spin rate of the planet. That is *not* a paradox. Perhaps the direct incoming sunlight could be blocked entirely; then the sunlight, to illuminate the planet, could be reflected from giant mirrors in 24-hour orbits *beyond* the shadows cast by the planetary sunshades. Such mirrors would rise and set with the cycle of the Sun as seen from Earth's surface, with one major difference: in order to remain clear of the planet's sunshade, the mirrors would have to be in a polar orbit whose plane was at right angles to the Sun-sunshade-Venus line. There are numerous variations on this idea which are yet to be worked out.

Conclusion

It is safe to say that the difficulties in terraforming Venus have been far greater than earlier theorists have liked to admit. However, we have also discovered that human ingenuity is probably far *more* powerful than we might have hoped. However mind-boggling the preceding suggestions may be, they must certainly be judged unimaginative in light of expected (but unpredictable) technologies for planetary engineering a century or two from now.

Our purpose in this chapter, therefore, has not been to raise insurmountable roadblocks to the terraforming of Venus, but rather to draw attention to the more serious difficulties in store for planetary engineers. In so doing, we can unleash the powers of human *imagination* against them. Facing such an unstoppable force, not even the hell-planet Venus can long endure.

The day Yahweh made the Earth and heavens, no plant of the field was yet in the Earth and no herb of the field had yet sprung up—for Yahweh had not caused it to rain upon the Earth. . . .

Yahwist Creation Cycle 9th century B.C.

10

The Smaller Worlds

It's good to be home, she thought as she watched the orbital shuttle disappear into the dark blue sky. *I'll never leave this world again—or want to.*

In a voyage symbolic of the yin and yang of the universe, the opposing forces and complementary actions, she had just returned from a duty tour on the rapidly-dwindling moon Amalthea. There, she had helped maintain the electromagnetic catapults which were daily propelling thousands of tons of the captured asteroid into far-flung orbits, creating a newer, denser ring around Jupiter. The world she had returned to, on the other hand, was being added to; five of Jupiter's outermost satellites had already been crashed onto its surface, and an entire equatorial ocean had been imported from Saturn.

For 800 years, the celestial object had been known as "Io," innermost of the Galilean satellites. But only a few years before, while she herself was still in Ho-Nan, the first surface dwellers had cracked their faceplates, breathed the still-thin air, and had shouted aloud the new name they had chosen for their new earth—"Epaphos," child of Io and Zeus in the Greek myths. It had been a fitting octocentennial celebration of human knowledge of the world.

Remaining at the landing field, she tasted the air again. *I swear I almost missed the sulfur during my absence. I know I did miss the crisp chill.* In exchange, she had given up, possibly forever, the sight of Jupiter filling half of the sky, its daytime pastels, and its nightime fireworks. The first settlements were on the far side of Io, where Jupiter never rose. *Three giant moons will be sufficient to light my skies,* she rationalized. *Oh, and the Sun too, of course,* paying brief homage to the source of her new world's daytime illumination, even as the planet's internal heat was the source of nine-tenths of its current warmth.

As if to close the last off-world chapter of her life, she deliberately turned towards the path leading to the commune, three miles down the creek. There would be a wagon along shortly but it would be busy enough with the luggage; maybe she could talk the tally-keeper into not charging her credit account for the ride she had decided to forego. *It will be good to walk along the path.*

In a larger sense, her path had been longer than the three final miles she was now walking. And that measurement was not in miles. She never thought of distance in terms of miles, only in time and cost—or even in maturation of mind. She still remembered her feelings of half-amused incredulity when she had first come upon the literature from the neo-Maoist syndicate, the pamphlets which called for a return to the ideals of traditional Chinese communism which had long ago been put aside. *Romantic reactionaries, that's what I thought of them.* She smiled as she noticed that Ganymede was rising.

They had been so dedicated, *pinpointing the first of their virtues which had favorably impressed her.* They knew what they wanted and they were willing to sacrifice present pleasures for future fulfillment. In late twenty-fourth century China, pleasures had been so inexpensive that many people dismissed them as worthless. Their counter-culture asceticism took many forms, but few *as potentially exciting as colonizing a new world.* Other parties had been impressed with the movement, and a committee in Hong Kong had convinced the authorities that this group would be a prime candidate for colonists once the bio-engineering of the southern hemisphere of Io was sufficiently advanced for surface settlement.

And I signed up before it was a reality, before it was anything but a promise, a hope she thought proudly. Today, of course, there were thousands of applicants for the hundred annual tickets and views of their activities were projected into every home in eastern Asia. There were still sacrifices and hardships ahead, but the conceptual breakthrough had been made. Io was habitable, and the communards had proved it.

A lifetime of plans lay ahead: the children she had scheduled, the new training in meteorology, all the books and cinema she wanted to absorb. And there were trees she wanted to watch grow, trees she had planted herself.

The underbrush along the path was thick but not high. There hasn't even been time for it to grow above my waist, *she knew.* I wonder what this path will be like when it's through a forest.

For a fleeting moment, she thought it might be too close, hemming her in. Had she made such a journey only to lock herself into a new pattern? She looked up. Europa was at the zenith, brilliantly blue-white as its frozen oceans and snowbanks reflected the distant sun. She had seen the plans for the giant space mirrors which someday would thaw those oceans and let the planet's stillborn life revive. What a pattern we could make on Europa, *she whispered to herself.* Easy, now. One step at a time.

Epaphos 2413 A.D.

Mars and Venus seem to us the most likely targets for terraforming because they already have atmospheres and because they are similar in size to Earth. As we have seen, both planets are indeed potential subjects for planetary engineering, but they have many problems, too.

There are other worlds in the Solar System, others besides Earth, Venus, and Mars. At first glance, such airless bodies as the Moon, Mercury, and Ganymede are far less likely candidates for terraforming efforts, but by now it's been well established that the first glance in terraforming is often misleading.

In fact, the more we look at the Moon, the more interesting it becomes. Transforming it into a more Earth-like world may not be as difficult as its presently foreboding appearance suggests. Furthermore, there are some surprising reasons which may make it useful to terraform the Moon within a century.

Other airless worlds present different problems and offer even less inducements for consideration for early terraforming. Nevertheless, each world will be discussed in its turn, and some surprising possibilities for their utilization can be proposed.

The first misconception to be discarded concerning the terraforming of the Moon is that the Moon's weak gravity could not hold onto the atmosphere once it had been installed. Over billions of years, or even over millions of years, it is true that such gases would be driven off by the Sun's heat. But we are speaking in terms of human history—in terms of thousands or tens of thousands of years, during which time continued human activities could maintain the atmosphere artificially.

Even with no outside intervention, such an atmosphere would be surprisingly long-lived. Writing in *Beyond Tomorrow*, asteroid colonization enthusiast Dandridge Cole asserted that "the lunar gravity would be sufficient to hold an atmosphere for several thousand years, so that would *not* be a problem."[1] A decade later, atmospherics physics expert Dr. Richard Vondrak of the Stanford Research Institute(now just "SRI") was asked what the dissipation rate of a 1 bar of oxygen-nitrogen atmosphere would be on the Moon. To drop only 10% he replied, would take thousands of years.[2]

Although the Moon at present has no atmosphere to speak of, there are ions(atoms stripped of their electrons) flying around near the Moon which could be considered a "mini-atmosphere." The total mass of this gas cloud does not exceed 10,000 kilograms. Thus, during an Apollo landing, for example, when tens of thousands of kilograms of burned propellants were being expelled by the lunar module engines, the mass of the lunar atmosphere was more than doubled—temporarily.

Dr. Richard Vondrak has studied the dynamics of the lunar atmosphere and points out that at present, the ions of the lunar atmosphere are essentially in free flight, impacting the lunar surface or escaping into space. There are so few of them that particle-particle collisions are rare. Within a short lifetime (measured in tens of days), the ion is driven by the energy of the solar wind to escape velocity or into the surface to combine with the regolith. New ions are constantly being generated from the surface rocks via ionization by the solar wind or by solar ultraviolet radiation.

When the Apollo missions added vast quantities of pollutants to the

The Moon's violent evolution has been entirely accidental. Now a gentle, deliberate evolution may remake it into a smaller version of Earth. *Art courtesy of Don Davis.*

lunar atmosphere, the normal sweeping mechanisms still worked well: after one month, only 10% of the propellants were left in the atmosphere, declining to 1% after another month.

But the sweeping out of the present lunar atmosphere by the solar wind is only effective as long as the wind has direct access to most of the atmospheric particles. A denser atmosphere would deflect the solar wind around the Moon. Says Vondrak, "Thermal escape then becomes the dominant loss mechanism and the atmosphere becomes long-lived, since thermal escape times are thousands of years for gases heavier than helium."

This semi-permanent atmosphere is still not particularly thick. Vondrak continues, "Although the transition to a long-lived atmosphere requires a total mass of 100 million kilograms, the surface density would still be low compared to densities in the terrestrial atmosphere."

Simple calculations would even tell us what mass of volatile material would be required to give the Moon a 1-bar (Earth-pressure) atmosphere. The Moon's lesser gravity would require six times the mass per unit area; the Moon's radius is one-fourth that of Earth, so its area is one-sixteenth that of Earth; the resulting mass of volatiles needed on the Moon is about one-third the amount of air needed on Earth. That's the equivalent of a sphere of water 60 kilometers across. In other words, a single medium-sized asteroid could provide enough volatile material to create a thick atmosphere on the Moon.

Vondrak has his own opinion about where to obtain such material.

> If one wanted intentionally to create an artificial lunar atmosphere, gases could be obtained by vaporization of the lunar soil. Approximately 25 MV is needed to produce one kg [kilogram] per second of oxygen by soil vaporization. An efficient mechanism for gas generation is subsurface mining with nuclear explosives. Krafft Ehricke estimates that a one kiloton nuclear device will form a cavern approximately 40 meters in diameter from which 10,000,000 kg of oxygen can be recovered. Application of this technique can easily generate the 100,000,000 kg of gas needed to drive the Moon into the long-lived atmosphere state.

But that would only be the beginning, because the air pressure would still be 10 billion times thinner than that on Earth. So back to the oxygen mines.

> An obvious speculation is the feasibility of creating artificial lunar atmosphere that would be "breathable" or as dense as the

surface terrestrial atmosphere. Such a lunar atmosphere would have a total mass of 2×10^{18} kg. Obtaining this much oxygen by vaporization of lunar soil requires an amount of energy equal to 2×10^{10} kilotons of TNT. Since this is approximately 10,000 times larger than the total U.S. stockpile of nuclear weapons, it seems impractical that such an amount of gas could be generated by current technology. Since there are no known natural gas reservoirs on the Moon, it would be necessary to import gases. For example, a cometary nucleus of radius 80 km contains 2×10^{18} kg of oxygen. [Vondrak concludes that] creation of a dense lunar atmosphere comparable to the terrestrial atmosphere is not feasible with current technology.

[Nor is it necessarily desirable, he cautions.] The desirability of intentionally increasing the density of the lunar atmosphere is highly questionable, since the primary applications of a lunar laboratory involve utilization of the present lunar "vacuum." But the artificial generation of an atmosphere can be considered as another potential method for modification of planetary environments.[3]

The Moon's Spin

Since we expect to import significant volumes of volatile material onto the lunar surface from deep space, we can also utilize the momentum of the incoming objects to spin up the Moon into a faster rotation rate.

Dandridge Cole first suggested this scenario in 1964, and he related how Russian theorists had also suggested the same thing. So obvious is the idea that when I began considering terraforming problems in 1977, it occurred to this writer as well; doubtlessly, many others have also thought of it. This writer's contribution to the scenario is to suggest that a stolen moon of Saturn be sent on a "reverse Jupiter swingby" into a retrograde orbit, thus increasing impact energy by a factor of ten. In fact, asteroids from the main belt can also be pushed *out* to Jupiter for the retrograde swingby, *saving* energy and increasing the payoff!

Despite some concern to the contrary, objects impacting on the Moon will not splatter out again into space, possibly smashing into Earth with significant destructive force. Dr. John O'Keefe of the NASA Goddard Space Center, in his attempt to explain the mystery of *tektites*, stated that the energy interchange dynamics of large impacts on the Moon almost insures that everything hitting the Moon stays there, and that pieces of the Moon are not going to be flung free like celestial buckshot to damage Earth.[4]

There will be problems in such a plan, of course. As pointed out by

The Smaller Worlds 227

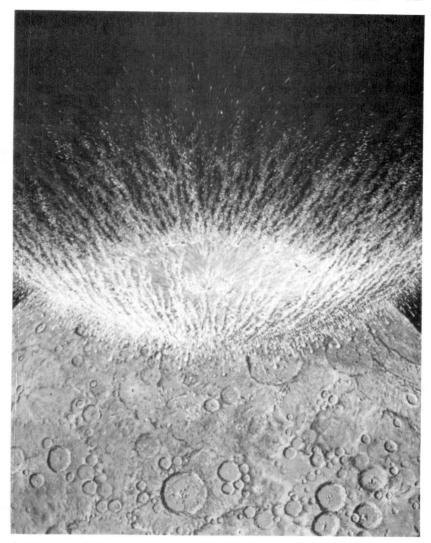

The Moon's gravity is strong enough to hold onto the impacting material during the initial explosion and for thousands of years afterwards.

John Hollenbeck, by the time we decide to import volatiles to the Moon, there will probably be an extensive human presence on the lunar surface in terms of factories, research stations, mines, and life support facilities.[5] The lunar population could be significant, numbering in the many thousands. Planned spin up impacts could do extensive damage to these facilities, perhaps requiring complete evacuation of the surface over a period of several years.

That depends on the mass of each individual iceteroid and on the dynamics of the interaction. If a very large number of quite small impacts are planned, the shock and the random fall of debris may not nearly be so dangerous. This would allow lunar inhabitants to go about their business almost as usual.

Suppose that we have emplaced a large volume of volatile materials on the lunar surface while at the same time succeeded in spinning up the small planet to a day-night cycle approximating that of Earth. What must we do next?

First consider that the *obliquity* (axis tilt) of the Moon's rotation is at present nearly zero. That is, the Moon spins about an axis which is almost perfectly at right angles to the plane of the orbit of the Earth-Moon system around the Sun. This means that there will be precious little seasonal variation on the Moon during a full year—no summers or winter, but just one continuous, eternal succession of days, each like all previous days and all future days.

Esthetics aside, could Earth biota adapt to such unchanging conditions? In the tropical regions on Earth they can and do, so a careful ecological plan could be drawn up. At least some variation of seasons would be desirable though, and the needed polar tilt could be implemented easily during the momentum input of the iceteroids. Do future inhabitable worlds need seasonal cycles? It's an excellent subject for ecological debate.

Lunar Seas

Now we have our 1-bar atmosphere, primarily water vapor and carbon dioxide and perhaps some ammonia and methane. The biological processing of that atmosphere into breathable air can be done as proposed for Mars and Venus. The Moon has the advantage that its surface temperatures would be very Earth-like once *any* air was in place.

Amazingly, we have only a vague idea about where the water will gather into seas, and where the true "maria" will form. For centuries, the darker regions on the face of the Moon have borne the names and titles of seas and oceans: Storms, Tranquility, Serenity, Clouds, Crises. They are, in fact, the lowlands since they were formed by flows of lava

The dry, misnamed "seas" may one day be covered with water; where once lava tubes formed river-like rilles on the surface, real water may flow.

from the Moon's interior—flows which did not manage to envelop the crater-covered highlands.

But even today, after Apollo, there do not exist planet-wide topographic maps of the Moon which show relative altitudes. Aboard two Apollo moonflights were laser altimeters, but the equipment was plagued with breakdowns and operational problems, and only a handful of complete orbits were ever charted. Extrapolations and good guesses can be made based on those few data points, but they are only guesses. The lower regions, the maria of the front side, are well identified; the higher

230   New Earths

regions, including frontside mountains and almost the entire backside, are also indicated. Yet many intermediate regions are not well plotted, and, with the budgetary strangulation of the unmanned *Prospector* (née Lunar Polar Orbiter) in 1978 and the subsequent cut off of Soviet space funds for lunar exploration, further such data will be long in coming.

The Moon's Winds Awake

Having estimated where the water of the Moon will gather, we next have to consider the wind systems—the driving forces behind the planetary climate. Here we must return to the principles described in earlier chapters concerning how a planetary biosphere functions. From those principles we can derive some plausible models for an inhabitable Moon.

The solar heating of the new Moon will be most intense around its equator, leading to rising masses of moist air (assuming we have equatorial open water, which we will, at least on what used to be called the "front" side). The moisutre subsequently condenses and causes violent rainfall.

As the rising air is replaced by colder air flowing toward the equator from both poles, it drops its water burden and moves in a contrary direction, northward in the northern hemisphere and southward in the southern hemisphere. On Earth, such poleward moving dry air masses eventually lose their temperature differentials with the lower layers of air and descend about one-third of the way to the poles, causing the worldwide belts of deserts. On the Moon, where the air will be much thicker (but no heavier, since gravity is one-sixth that of Earth) and the equator-

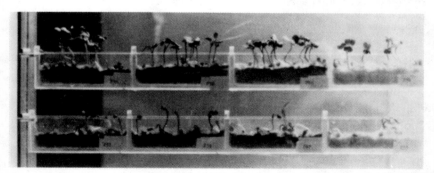

Tests on Earth showed that plants would grow in Moon soil, with some additional material supplied.

to-pole distance four times less, we can suspect that the upper layers of the atmosphere will *not* descend at some intermediate point but will travel all the way to the poles prior to sinking and creating deserts.

Planetary atmospheric circulation will then be essentially one circulation cell per hemisphere. The polar regions will be deserts since the dominant air masses will be dry and cold. From the polar regions, winds will fan out towards the equator, veering more and more to the west as the "Coriolis effect" grows. Instead of bands of easterlies and westerlies as on Earth, the Moon's wind patterns may be far simpler: northerly winds in the upper latitudes (using the northern hemisphere as an example), shifting gradually to north-easterly winds nearer the equator. And depending on the placement of the bodies of open water, rainfall patterns will also be predictable.

Why Terraform the Moon?

Because of the Moon's proximity to Earth, it should be considered as an early terraforming project, despite the fact that it has no atmosphere at all (since in Venus' case, that atmosphere is a definite *drawback,* and in the case of Mars, some additional volatiles may need to be imported as well). The level of solar heating on the Moon is, quite obviously, the same as that on Earth, and that may simplify the problems.

But other open-minded futurists still questioned the practicality of making such a prodigious effort for uncertain benefits, at least in the case of the Moon. Dandridge Cole, who was a leading advocate of colonizing the asteroid belt, was critical of schemes for terraforming the Moon. Writing in *Beyond Tomorrow,* Cole claimed: "Terraforming the moon would be a nostalgic attempt to return to the primitive open cycle environment of pretechnological Earth. It would be a fantastically inefficient process in that thousands of times more oxygen would be produced than could be used by the early colonists. By the time billions of inhabitants of the Moon were making efficient use and reuse of the atmospheric oxygen, they would necessarily have had to close the environmental cycle anyway. Why not start with efficient closed cycles since that will be the inevitable end result?"*

But one very practical and very persuasive argument for terraforming the Moon has still not been appreciated. There may be excellent reasons to emplace an artificial atmosphere on the Moon before the end of the

---
*Cole's untimely death on October 29, 1965, at the age of 44, robbed the future of a farsighted, eloquent and level-headed spokesman.

twentieth century, for the benefit of Earth and without regard to what actually happens on the lunar surface.

The justification for such a project stems from ex-astronaut Brian O'Leary's recent proposals for the retrieval and industrial exploitation of passing asteroids. Such a project, O'Leary calculates, could produce trillions of dollars worth of metallic ores, while only costing in the tens of billions of dollars.

The asteroids would be fetched back to orbit above the Earth using mass drivers which could eject part of the asteroid's own body as reaction mass, pushing the remainder in the desired direction. It could take months or years to nudge the asteroids onto courses towards Earth, so O'Leary suggests that these missions be controlled by sophisticated robot pilots. One problem for such a scenario is the fact that the asteroids approach Earth on a relatively rapid trajectory, with speeds which must be decreased quickly if the asteroid is to be captured for a stable orbit within range of the waiting space transports. The speeds it took the mass drivers years to build up would now have to be reduced within hours.

But the technique of *atmospheric graze* could be useful here. By diverting the asteroid through the outer fringes of a planetary atmosphere, most of its velocity could be sharply diminished at no cost in propulsion expenditures. But it is inconceivable that such maneuvers could be permitted to use *Earth's* atmosphere because of the possibility of hemisphere-wide devastation if the asteroid's flight path is off by only a few kilometers.

The alternative, offered by this writer, is to plan such an atmospheric graze maneuver using the atmosphere of the Moon—a thin atmosphere emplaced primarily for this use as a brake on incoming asteroids. There would be no danger to human life, but the payoff would be dramatic by decreasing the cost of asteroid retrieval missions significantly. And another problem in space propulsion would be eliminated: even if the dangerous Earth-atmosphere graze maneuver were somehow approved, the result would be the creation of asteroid trajectories which would reencounter Earth within a few weeks after the first graze unless some other propulsion system were urgently used to trim the orbit into a safe one. Since the Moon circles Earth, an asteroid which grazed the Moon's atmosphere and subsequently returned to the same spot a few weeks later would find that the Moon was no longer there. The asteroid's orbit would be safe with a *single* encounter, and no crucial follow-up maneuvers would be demanded to avoid planetary disaster.

Scenario for a New Moon

The Moon will not long remain without an atmosphere, whatever may be the ultimate intentions of human spacefarers. There has always been some gas around the Moon, temporarily trapped from the solar wind, baked from the lunar soil, or gently belched from the banked fires of the deep lunar interior. Our Apollo visits to the Moon each temporarily doubled the mass of this lunar "atmosphere," and more frequent return expeditions in larger space ships will contribute even more.

All of this material is invisible from Earth. However, if proposals for installing an "asteroid speed brake" in the form of a substantial atmosphere are realized, or if the Moon is terraformed for other reasons, there will be tremendous changes in the aspect of the Moon as seen from Earth.

First, of course, it will be brighter, perhaps five times brighter or more. "The lesser light which rules the night" will cast even more powerful shadows when the Sun has set. Additionally, the specular reflection of the Sun off the near-side lunar oceans will occasionally present a dazzling point source of light on the visible face of the Moon which will last for several hours as the Moon and Earth move into and out of the appropriate alignment with the Sun. Less dazzling but more commonplace will be the specular reflection of the Earth itself off the lunar oceans—a reflection which should be visible all across the night side of Earth. Astronomers will notice that occultations of stars will no longer be abrupt, but will consist of a gentle dimming decorated by soft color changes in the starlight. The new Moon envisaged here will cast a new spell in the night skies of old Earth.

Mercury in its Flight

Obviously, sunlight is much too bright on present-day Mercury. Less obviously, sunlight is much too irregular to be harnessed even by the "simple" techniques described in chapter 8.

At its closest point to the Sun—its *perihelion*—Mercury receives more than ten times as much solar energy as does Earth. But due to the extreme eccentricity of the planet's lop-sided orbit, the sunlight falling on Mercury six weeks later (at *aphelion,* the farthest point from the Sun) is only half as great as that during the close pass. This pattern is repeated every 88 days for the whole planet; the rotation rate of 59 days gives Mercury a sunrise-to-sunrise "day" of 176 Earth-days which results in

Mercury lies sun-baked and sterile, but not for long.

one side of Mercury being exposed to the perihelion sunlight and the other side being exposed to the half-as-strong aphelion sunlight. The geologic implications of this are not well understood; the terraforming implications are that half of Mercury might be made habitable while the other half remained too hot.

Mercury has one-seventh the land surface of Earth and 37% of the surface gravity. This means that its atmosphere mass would have to be about 40% that of Earth's to provide Earth-normal surface pressure (about the same as needed on the Moon which is the equivalent of an iceteroid 60 kilometers across). The atmosphere would have one immediate effect: it would probably increase the planet's albedo from the present 0.07 to a figure approaching 0.60, thus effectively cutting in half the amount of solar energy absorbed by the surface (from 93% to 40%). Properly mixed from imported volatiles, the atmosphere could have a minimal "greenhouse effect" (nothing more would be desirable so close to the Sun).

Techniques already mentioned, such as giant parasols, artificial rings, or shadow-casting dust clouds, could help reduce solar heating of Mercury. Some sophistication is called for in designing a sun-shield that can accommodate the factor of two changes in heat over the 88-day year.

The problems of sunlight, of building an atmosphere and hydrosphere, and of tolerating the intense radiation from solar storms, all seem to indicate that terraforming Mercury will be put off for a long period of time. The major scale of such planetary engineering suggests that extremely powerful tactics can be required for Mercury, such as the following: its orbit can be smoothed out, and needed volatiles can be imported all by the single expediency of colliding Jupiter's ex-moon Callisto into Mercury. As we'll see in a moment, Callisto may not be good for much else anyway.

Ultimately, Mercury can be moved out from the Sun, perhaps into an orbit around Venus. Such massive efforts are today almost unimaginable, but in Mercury's case, not much less will do.

The Galilean Worlds

Early in 1979 we discovered four new worlds. True, their existence had been known since 1610 and their motions had been plotted with great precision for centuries. True, observations from Earth and by earlier space probes had uncovered hints about their natures. But when Voyager 1 passed Io, Ganymede, Europa and Callisto in March, and Voyager 2 probed the same system in July, photographing their surfaces and discovering hitherto unknown physical processes, four blank places on our universe maps were filled in.

It is natural that we consider those worlds as candidates for terraforming, both for nostalgic grounds (Ganymede was the stage for *Farmer in the Sky*) and practical grounds (the planets are the size of the Moon and Mercury). But several immediate difficulties arise: first, the worlds are surrounded by lethal radiation trapped in Jupiter's magnetic field; second, the worlds receive less than one-twenty-fifth as much solar energy as reaches Earth; third, one of the worlds seems to have no water at all while two others would (if warmed) have oceans 1,000 kilometers deep. Is there any conceivable way to overcome these problems?

Taking the easiest problem first, the low level of sunlight can probably be approached the same way it was over Mars, only more so—with mirrors. To increase the sunlight hitting each planet by a factor of twenty-five, mirror systems with a total diameter of about five times the planet's own diameter would be needed. The engineering aspects of such a system (including gravitational perturbations by Jupiter) are formidable; the conceptual aspects present no problem.

Facing the radiation belt question squarely, a hint of a solution was accidentally dropped during the debate over the newly discovered rings of Jupiter. One observer suggested that earlier Pioneer probes had made readings which should have suggested the presence of a ring system: the radiation readings had dropped momentarily, as it turned out, due to a sweeping up of the energetic particles by the material of the ring.

A larger artificial ring system could then increase this effect and sweep up much of the radiation; to create it, portions of the inner moon Amalthea (or smaller nearby moonlets) could be disintegrated. In collusion with this approach, a direct assault on the Jovian magnetic field which induces the radiation belts could also be launched. Such a project could make use of the inadvertent (and nearly disastrous) results of high-altitude nuclear bursts conducted under Project Argus and Project Starfish in the period 1961–1963. These blasts, at altitudes of 100 kilometers and higher, injected quantities of radiation into Earth's van Allen belts, but they also caused a "dumping" of charged particles into the belts, resulting in some cases in a lower level of radiation because of the overloading which caused the belts to collapse. Lastly, streams of ions could be artificially generated, much as is done naturally by the volcanoes of Io, to "short circuit" the electromagnetic field of Jupiter and weaken its ability to hold on to charged particles.

Io: Fire, But No Ice

The surprise of finding almost a dozen active volcanoes and nearly fifty active gas vents on Io, innermost of the four Galilean moons of Jupiter, was probably matched only by the satisfaction of three space scientists who had predicted that such activity probably existed, caused by tidal heating of Io by Jupiter. That is, the gravitational stresses in the crust of the Moon-sized world heat the rock enough to expel any meltable material through the surface into space: thus had written Dr. S.J. Teale, Dr. P. Cassen, and Dr. R.T. Reynolds.[6]

That appears to be the pattern on Io, since the active volcanoes and the dozens of extinct volcanic calderas seem to cluster around the equator, where tidal heating would be most severe. Some of the expelled material is sprayed into space, but most of it falls back to the surface where it buries the meteor craters which must have once formed on Io. Yet not a single crater has been identified in Voyager photographs, suggesting that the entire crust is resurfaced by the volcanoes on a time scale of less than a million years, perhaps much less. As new layers form, the more deeply buried layers are warmed by the tidal heating until they melt and spray out again.

The material involved is probably sulfur; no trace of water or carbon dioxide (characteristic of Earth volcanoes) has been found in the volcanic plumes. Sulfur in its various chemical forms can assume a wide variety of colors, and could by itself account for the striking reds and blues and whites and blacks of the rainbow face of Io.

At the poles of Io are mountainous areas, up to 10 kilometers high, which may be remnants of Io's primitive crust formed before the tidal heating began fueling the sulfur volcanoes. Elsewhere the planet is very, very flat.

The prospects for terraforming Io are uncertain at best. The volume of expelled sulfur is not high, so a newly introduced thick atmosphere would probably be able to swallow the ongoing contributions of the volcanoes, except that the chemistry would lead to sulfur dioxide being formed, leading to sulfuric acid in local rainfall. Even less absorbable would be the sulfur already coating the surface, which would probably poison any bodies of open water which would collect. Before Io is terraformed, it may be necessary to bury the present surface beneath several kilometers of rock imported from one or more of Jupiter's outer moons, or from nearby Amalthea.

On the other hand, Io has the advantage of having its own internal heat source, compensating for its great distance from the Sun. Also, its "day" is not much different from Earth's and could be made more similiar by impacting incoming materials at such an angle as to create spin up. Upon closer consideration, Io becomes more and more attractive as a terraforming candidate.

Europa: The Frozen Flood

Europa, next Galilean world out from Io, has a lower density than

Io and consequently seems to have large amounts of water already in place while Io has none. In fact, Europa may have too much water already.

A large fraction of the bulk composition of this world may be water-ice mixed with rock and dirt. Based on observations which show long intersecting linear features suggestive of continental drift, scientists now suspect that the rocky core of Europa may be overlaid by a layer of water-ice up to 100 kilometers thick. Melted, such a layer would provide an ocean several times too massive, drowning all dry land.

Potentially, however, the excess water could be exported for use elsewhere, or simply be boiled away by the input of concentrated solar or nuclear energy. While this was in process, the heated new oceans could be seeded with aquatic life which could begin the atmospheric transformation even before the dry land appeared. On Europa, it would be up to human efforts to first melt the waters and then remove them as they expectantly watched for the new Ionian continents to rise from the sea!

Ganymede—Ocean Planet or Planet Ocean?

The Voyager photographs were most revealing about the surface of Ganymede, third of the Galilean worlds. The pictures showed both heavily cratered regions and a hitherto-unimagined type of surface now called "grooved terrain." These discoveries held great significance for scientists attempting to fathom the history of the Solar System.

For would-be terraformers, the news was that the density of Ganymede was confirmed to be so light that at least half the world's mass is probably water-ice. Structurally, this indicates that a small rocky core is surrounded by a 1,000-kilometer-deep layer of water, the outer surface of which is frozen, but the mass of which is probably liquid (which might cause one to speculate about the kind of habitat this would provide for native life forms). The weakness of the icy crust of Ganymede was underscored by the observation that the surface was utterly flat, with no variations greater than a few hundred meters across the entire face of the world.

So what is Ganymede good for? Warming it would only produce 1,000 kilometers of ocean. The energy to remove that water would be prodigious, and the resulting stony world would be smaller than Earth's own Moon.

Ganymede may be just the thing for implementing the Burns-Harwit Maneuver over Mars. Ganymede would in this case be transferred to Mars via an economical reverse Saturn fly-by (being careful to minimize gravitational disruption of the ring system), then spend several decades making close passages to the Sun to boil off the excess water (forming by far the most spectacular comet the Solar System has seen in eons). It would then make several Mars fly-by maneuvers to help even out the orbit of Mars and place Ganymede on a new path close enough to Mars so that capture could (with some assistance) occur. By the time it settled into its new orbit, Ganymede could be another habitable world.

Since Ganymede in mythology was said to be the male counterpart of Hebe, the goddess of freshness and youth in nature, the transformed moon Ganymede might deserve such a new name—Hebe. That name already belongs to a small asteroid, but after several centuries of human engineering activity in the Solar System, the asteroid may no longer exist, and a new world may have need of that name!

Callisto—Stealing the Stolen Nymph

Much the same problem (and opportunity) arises with Callisto, last of the four Galilean worlds. Its density is very similar to that of Ganymede, signifying that a warmed Callisto would also have oceans many hundreds of kilometers deep. Its present surface is extremely flat and crater-scarred, with the traces of giant impact basins, but with all mountains and valleys smoothed out by the elastic flow of the ice crust over the eons. There are no grooves on Callisto, but the surface is considerably darker than the ice surface of Ganymede, suggesting that some layer of dirt has been mixed in.

Callisto, too, may ultimately be torn from Jupiter's grasp for terraforming purposes. This would only be fair, since in mythology Jupiter had originally seduced her, fathering her child Arcas, first king of Arcadia. Arcadia in turn is the traditional birthplace of Hermes, known to the Romans as Mercury. This ancient myth may demonstrate uncanny prescience regarding the ultimate fate of Callisto, the identity of the father of her last planet-child, and even the best name for that new Earth.

If Callisto has no value for terraforming while it still circles Jupiter, then it can be of great value elsewhere. Perhaps Callisto can be put into an orbit close to Earth, after its excess waters have been boiled off by several dives through the Sun's flames. Perhaps instead, the water and

One solution might be to combine one too-wet world with one too-dry world, to form a medium-sized planet with ample oceans and air. Here artist Pat Rawlings portrays the engineered collision of Callisto and Mercury.

the momentum of Callisto can be utilized on another world which needs such oceans and needs such momentum to give it a shove away from the Sun: that is, Mercury.

Here is a suggested scenario. By smashing Callisto into Mercury, the combined new world (and why *not* call it Arcadia?) would be the size of Mars with plentiful oceans and an orbit taking it near Venus. There it could be gently nudged into orbit around Venus where its gravity could help churn the iron core of Earth's twin, generating magnetic fields needed to complete the duplication of Earth (and the gravity of Venus would do the same favor for the iron core of Arcadia). Then the scorching regions closer to the Sun could henceforth be reserved for suicidal comets and off course asteroids such as the aptly named Icarus; it would be no place for new Earths.

Beyond the Galilean Worlds

Beyond Jupiter, the rocks grow scarcer or are more deeply buried.

Titan and Triton are moons nearly as large as Mars, but are even more predominantly water-ice than are the challenging worlds of Ganymede and Callisto. If we seek rocky cores for additional new Earths, we must go into the hearts of the gas giants themselves.

Neptune and Uranus could be the first candidates to follow Ganymede on its cleansing dive past the Sun; each planet probably has enough stony material to form a planet perhaps the size of Mars (or several the size of the Moon) once the thousands of kilometers of water, ammonia, and methane are boiled off. Unfortunately, the interiors of Uranus and Neptune are probably already extremely hot due to the present gravitational pressures on them. Boiling off most of the planetary mass by skimming the surface of the Sun would just add to that heat. The result would be that although the stony cores of Uranus and Neptune could be exposed, the new planets would consist of molten rock which could take a million years to cool naturally. Techniques effective in freezing solid the molten lava covering an entire Earth-sized planet have not been invented yet!

The problems are if anything even more insuperable in the case of extracting the rocky cores of Saturn and Jupiter. It might be far easier to return to the inner Solar System and consider dismantling Venus and Earth, creating in place of each planet a brood of eighty Moon-sized planets whose total surface area would be five times as great as that of each single parent. I leave such speculations to the *second* half of the next millennium.

> *Space is huge enough, so that somewhere in its vastness there will always be a place for rebels and outlaws. . . . The open frontier will beckon as it has beckoned before, to persecuted minorities escaping from oppression, to religious fanatics escaping from their neighbors, to recalcitrant teenagers escaping from their parents, to lovers of solitude escaping from crowds. Perhaps most important of all for Man's future, there will be groups of people setting out to find a place where they can be free from prying eyes, free to experiment with the creation of radically new types of human beings, surpassing us in mental capacities as we surpass the apes.*
>
> Freeman Dyson

11

Why? Who? By What Right?

Closest passage was still hours away when Weaver drifted into the transparent half-bubble which occupied the twilight facet of his family's lodge. He adjusted the alignment of two of the membrane screens and, satisfied with the resulting privacy, deployed his hammock between them.

No, I'll never be ungrateful for such a view of the stars, he realized, staring past the looming globe already too large to cover with the hand of an outstretched arm. But it is exciting to see a planet so close. He anticipated the coming delicious terror of falling—a sensation rarely felt by those born, bred, and raised in free-fall between worlds. Someday our lodge may raise a child to walk the future surface of that earth.

The cloud-shrouded world was sunlit only over a narrow crescent, since Weaver's lodge was overtaking it from downsun. Across the daylit side no features could be discerned in the nearly-blinding glare, but the night side—"night" being an archaic term Weaver had learned at school—had dim sources of flickering illumination: violent aurorae, constant lightning storms, longitudinal bands of atmospheric chemoluminescence, and localized volcanism. A dozen blue-green beacons shown steadily: the megalasers, pumping thermal energy from the surface into deep space, a millionth of a degree at a time. Weaver shuddered. Such awesome untamed energies. How—how uncivilized.

To the eye, the star-filled vista beyond Arkady was silent and static, but Weaver knew better. The spaces before him were abuzz with communications and aflame with artificial energies. Several inhabited planets were in view; three other planets had dim, cometary tails—excess hydrogen being boiled off the fragmented cores of the former trans-Saturnian worlds. Around the settled planets—old Earth itself was not in view—were clusters of lights, each world a jewel set amidst a miniature human-made Pleiades. The lights showed the dense pattern of artificial islands holding hundreds of millions of people.

A century before, when the planet now called Arkady was constructed from a guided collision between Mercury and Callisto, Weaver's family had been prospecting some likely-looking asteroids on the outer edge of the Belt, some of which had been perturbed into new orbits by the final close passages of Callisto between its theft from Jupiter and its marriage with Mercury. The lodge had come across another silent lodge, broken and burrowed into a small asteroid. Its inhabitants—all three hundred—had been dead for centuries. According to records they'd found aboard, most of the miners had been earthborn emigrants—and the transition to deepspacing had been too great a psychic leap for their instincts and skills.

Among the frozen bodies were two dozen children and infants. Deeply moved, Weaver's family had adopted the dead young and had cloned embryos of their own from undamaged cells. Weaver himself was one of those reborns.

Genetically that makes me a first-generation spacefarer, he had often thought to himself. That had never meant anything in behavioral terms, until this present moment. He looked again at Arkady, visibly bigger, with stars fading out along the nightside limb and others emerging from the glare of the day side. The lodge would be zooming past at a range of two planetary diameters before this watch was over.

There's something about a planet . . . he realized. Five billion years of evolution count for something. He thought about real wind on the face, not an air duct current; about real rain on the back, not a shower;

the scent of ozone in the air after a thunderstorm, signifying the passage of danger even while ozone in a lodge was the sign of danger's approach; real heat and cold, sounds and silence, all uncontrolled; real gravity. And an unlimited number of people to meet. Lodges rarely met except on asteroids being mined by rival—and hence usually unfriendly —combines, and although there were always a handful of other lodges passing within radioconversational range (a few million miles at most—the delay lag became too distracting beyond that) such intercourse offered no physical contact. Perhaps we can all meet on Arkady.

This body could feel at home on a planet, Weaver thought, distracted momentarily by the arrival of siblings in the viewing room (he could tell where they had been at school—they were still chattering in Napoleonic French). The world's not for me—I haven't the muscles of a mudfoot, or the reflexes—but my next double, he or she should be the one.

He had seen all he wanted of Arkady, and unhooked his hammock, leaving room for others at the window. The view—and his vow—would remain with him for the rest of his life.

Above Arkady, 2771 A.D.

The motivations for rebuilding planets may turn out to be not as esoteric as the exotic nature of the project might suggest. It is, after all, only a larger scale of environmental engineering, along with such precedents as swamp draining, forest clearing, desert irrigation, and buffalo killing—all situations in which the natural order of parts of the universe have been intentionally destroyed in order to provide living space for people. The same practical justifications may be given for terraforming; the same moral arguments may be raised in opposition.

Some insight into the processes which might lead to a decision to conduct a terraforming project can be found in the transcripts of a special session of the American Association for the Advancement of Science, published in 1978. The topic of the meeting held in Chicago in January 1978 was *macro-engineering,* a new word coined to describe engineering projects of great magnitude, past, present, and future.[1]

Discussing such historical perspectives, Dr. Eugene Ferguson of the

University of Alabama defined the term. "A macro-engineering project in any age is one that strains the current capabilities and resources. It is at the outer limits of the 'state of the art'; it is expensive; and because it is so different from run-of-the-mill projects, the macro-engineering project [or "MEP"] furnishes engineers and technicians with a particular challenge and fascination."

Such projects are not new. The earliest-remembered MEP, according to Ferguson, was the Tower of Babel, which has come to typify the entire mentality behind MEPs. "It has become a symbol of what happens when audacious men push their enthusiasms to a point where they challenge the order of a God-centered world."

Making over another world, from the form "in which God created it" into a form suitable for human needs, is certainly susceptible to this kind of criticism. This is a warning that the issues could be more than just environmental and practical; essential concepts of philosophy and theology may come into play.

Another MEP was the Great Pyramid of Giza, built by engineering descendants of Imhotep, the world's first great engineer (he created the stepped pyramid at Saqqara nearly 5,000 years ago, and was deified in Egyptian mythology for it).

Similarly, the Great Wall of China, probably built mainly within thirty years of activity interspersed over a 200-year period around 200 B.C., was conceived and designed by one man, the engineer Ch'in Shih Huang Ti. At least one million men must have been employed in the total construction, but it was built in a time of peace when the Emperor had enormous teams of unemployed soldiers at his disposal.

The Suez Canal is another ancient MEP, having been built and rebuilt for millennia. An inscription at Karnak suggests that it was being used in 1380 B.C.; it was repaired by Roman emperors in the first century A.D. but closed down by the Islamic Crusades around 770 A.D.; the Arabs opened it again briefly around the year 1,000; French engineers built it one more time in 1869.

Ferguson puts his finger on another vital aspect of the probable course towards terraforming. "An MEP is the result of a visionary, a person of energy and imagination caught up in his own enthusiasm or those of his times." No great accomplishments ever seem to come about because some committee voted on them: it took a single person with unique characteristics, acting at a fluid moment of history, to swing the course of a civilization from one direction to another, to channel its energies into a monumental project. There is no reason to suspect that it will be any different when the time comes to submit the budget for rebuilding Mars. One person will champion the initial commitment to the project.

This fits in well with another aspect of these colossal projects: they are not "useful" in a materialistic sense. The motives of the people behind such projects, writes Ferguson, "are more aesthetic and psychological than economic and practical. In general, an MEP usually does not satisfy some simple specifiable immediate social need (although it may, sometimes). The needs it does satisfy are often if not always on a different plane: spiritual, psychological, aesthetic, patriotic."

Such an observation may help overcome the major argument heard against terraforming, an argument invariably resorted to after all previous objections to the possibility and feasibility of such projects have been overcome: "Why take the trouble?" Lists of possible purposes can be made up, and they can be referred to, but usually without effect.

Instead, the attitude of would-be terraformers can be much more optimistic and aggressive. If terraforming requires a rational and budget-conscious basis, it will be the first such giant project ever to require such. Dozens of earlier projects, each as grossly expensive in relative terms, have been tackled, and in some cases we see the results but still cannot begin to guess "why?"

Ferguson gives other valuable cautions. "A little reflection tells us that engineers have never seriously considered the social costs of their projects. . . .clearly we need nontechnical people—responsible people who do not share the presuppositions of engineers—to be actively engaged in the planning stages of a macro-project, so that the social costs will not be overlooked as projectors go on to bigger and more exciting things." And such a cautionary approach should not overlook consultations with noncommitted engineers as well, since "because of the enthusiasms that created and sustain it, an MEP may have grave technical faults."

These well-taken warnings, which today can be applied to new canals, dams, supersonic transports, spaceships, skyscrapers, and power plants of various species, must go double for future consideration of planetary engineering on Earth or on other worlds. The greater the enthusiasms, the greater the risks of fatal flaws in the plans. That has been the lesson of the history of MEPs.

Parallels for Terraforming

In approaching this question of "why?" from a rational (or rationalizing?) point of view, numerous types of answers are possible. Generally they rely on analogies to human colonization activities of the past, primarily Western activities. Granted that such a reliance is bound to result in an incomplete picture, the view is still interesting and is probably quite helpful.

For reference, these different theories can be labeled with some mnemonic tag signifying the essence of the argument. I call these theories the "Everest Syndrome," the "Eldorado Syndrome," the "Tobacco Syndrome," the "Botany Bay Syndrome," the "Mayflower Syndrome" (with its variant subclass the "Erik-the-Red Syndrome"), the "Unemployed Soldiers Syndrome," the "Show-the-Flag Syndrome," and the "Wild West Syndrome." Each has its own insights to give, although their applicability may vary widely.

The "Everest Syndrome" is, of course, based on Sir Edmund Hillary's fatuous reply to the equally fatuous question, "why climb that mountain?" Either in an attempt at humor, or to put the reporter off, or to hint at philosophical profundities (or all of the above), Sir Edmund replied simply, "Because it's there."

Yet the answer is not nearly as trivial as it might seem. It calls up a basic human drive, the desire to see over the next mountain and, while looking, to scratch one's initials on the side of the highest rock. Such a drive is a valid one, long established and recognized. It is a good excuse for individual or small team exploration, or for the recruitment of colonists, but it seems to be a poor motivational factor to use in enlisting the support of an entire society.

The "Eldorado Syndrome" refers to the search for valuable materials, either raw resources or ready-made kingdoms ripe for looting. Such a justification can be applied to the exploration of the Solar System, even to the level of extensive human participation in interplanetary travel. But its application to planetary engineering is tenuous, unless it could be demonstrated that terraforming would facilitate the search on a planet. This could come about by causing deep excavation or extensive erosion to uncover the resource, or by activation of some new conveniences to support human activities. The emplacement of an atmosphere and hydrosphere on a planet could greatly facilitate point-to-point transportation by allowing the introduction of such time-tested transport techniques as winged aircraft, canoes, and burros.

The "Tobacco Syndrome" is what made the Virginia colonies prosper when the adventurers had given up looking for gold. As it turned out, the climatic and soil conditions in the colony were such that agriculture would produce a luxury item high in demand in the homeland (the same function later provided by the fur trade). Some exotic crop—perhaps a drug, a spice, or a base for a perfume or a liquor—could be grown on a terraformed world but not on Earth. The market for Martian truffles, Rigelian brandy, or Mercurian marijuana is unpredictable but potentially vast.

Yet it is hard to imagine that such products could not be produced as well in close-Earth space colonies where planetary conditions can be duplicated. However, products such as these have been the wild-card surprises of colonization efforts in the past, and they may surprise us again.

The "Botany Bay Syndrome" is a potential payoff for terraforming, since it takes a page from Australian history in an era when Britain was sending off its "undesirables" on one-way journeys to a wasteland. The result is a prosperous and happy nation of people who still to this day often ask themselves what they are doing on the wrong side of Earth 20,000 kilometers from their Mother Country. But measured in travel time and expense, a terraformed Mars of the late twenty-first century would be no more distant from Earth than was the mid-nineteenth century Australia from Britain.

The "Mayflower Syndrome" refers to the Pilgrims who settled in Massachusetts in the early seventeenth century. This event has become shrouded in myth and misinformation over the centuries, but a better examination of it could yield valuable insights for space colonization.

First of all, as Freeman Dyson loves to point out, the voyage was not cheap; the would-be colonists did not build a boat and sail blithely off into the sunset. Instead, they wheeled and dealed to obtain backing from London financiers, and they individually put up what would be the equivalent of a twentieth century person's life savings.

And there was something special about the colonists themselves: they were not a slapdash conglomeration of recruits but were instead a close-knit religious community that had been together for more than ten years. Sociologist Paul Meadows quoted a study of utopian communities (a study which has much to teach would-be space colony sociologists and socio-engineers) pointing out that for such groups the physical separation from the parent society was only the last step in the formation of their mini-society: a society which had been congealing and distinguishing itself from its parent culture even as the members lived within it.

Lastly, they were not actively seeking a frontier life: it was forced upon them as their only choice to maintain their mini-society intact from outside interference. They were forced to the frontier.

The bottom line here is that this "Mayflower Syndrome" portrays a colonization effort in which the colonists are not at all representative of their parent culture: they are, bluntly, deviants from the norm at home. While this may seem at first a major impediment to obtaining the necessary economic and technological support from the home culture, in fact just the opposite may be true: the home culture may be willing to expend a great deal of money and effort to get rid of its social deviants.

A variant of this theme is the "Erik-the-Red Syndrome," named for the father of Leif Erikson who settled in Iceland in the ninth century A.D. Iceland itself was a rough-and-tumble pioneer community, but Erik still got himself into trouble with what passed for the law. Fleeing with his followers, he settled on the southwest coast of Greenland (so-named because if he had named it what it really looked like, nobody would have followed him) and founded a relatively prosperous society. His followers chose exile rather than face staying home.*

The "Unemployed Soldiers Syndrome" refers to a government's peaceful use of military forces which due to lack of overt warfare are not engaged in combat. As an alternative to demobilization (perhaps in expectation of future hostilities or to avoid massive economic dislocations), the soldiers and the war machines are turned to exploration, colonization, or massive public works projects.

Numerous examples can be found in history, from the Roman legionaires being settled in Dacia (modern Romania) and the Cossack "mir" settlements on Russia's expanding frontier in the period 1600–1900, to the US Navy's expeditions to Antarctica in the late 1940s and the US Air Force's support of America's space exploration program in the first decade of the Space Age.

This not-altogether-pleasant scenario deals with the regrettable fact

---

*The Greenland colony ultimately came to grief when Earth's climate changed for the worse, wiping the colony out in the early fourteenth century. Some local climate modifications might have saved them. If a similar predicament threatens Earth again, it will be time for terraforming to come to the rescue here on our own home planet.

that many of the tools of terraforming (such as high-energy lasers, thermonuclear explosives, giant mirrors, and other planet-moving and atmosphere-modifying techniques) could be modified for military purposes. Hopefully the reverse of this, the "swords-into-plowshares" transformation will be more likely: giant space weapons of the twenty-first century, never used in anger, might provide readymade war surplus terraforming machines after the political tension which begat them had passed. It has happened that way before.

The "Show-the-Flag" syndrome is an unabashed appeal to national chauvinism and xenophobia, but its effectiveness cannot be gainsaid since the results of such competitive scrambles are usually beneficial long after the sordid rationalizations for their initiation have passed. Perhaps the best example is the seventeenth century history of the colonization of the east coast of North America, when numerous European nation-states each tried to stake out their claims, usually in response to perceived threats from other nation-states. Britain, France, Holland, Sweden, Spain, and other parties all devoted considerable energies to establishing settlements. As a result, all ultimately contributed to the vigor and prosperity of the United States and Canada.

The "Wild West Syndrome" is named after the American frontier of the nineteenth century. To be sure, the frontier was a place where the adventurous, the desperate, the ambitious, the deviant, and the fugitive could find refuge. This is the "escape hatch" effect, which promised "open skies" to those who felt hemmed in by neighbors and family, by social conventions, or by the physical pressure of crowded cities. But beyond that, just the *idea* of the frontier, as exemplified by the contemporary myths in the dime novels, invigorated the stay-at-homes with thrilling stories of adventure, courage, hardship, and triumph. Those who settled the West were for the most part a representative sample of the native culture, with which the stay-at-homes could successfully identify and vicariously experience the frontier mentality. All in all, it was an exciting era which stamped itself upon America's national consciousness. A century later, America has not reconciled itself to the demise of the Western frontier.

With the construction and settlements of new planets, we can reproduce much of that era with all the mysteries and challenges and hard-fought battles with nature and with all its stimuli to the national self image. For "national," read "global": the Japanese, Italians, Germans and others have adapted the American Western myths with a vengeance, and other national myths such as the Russian, the Brazilian, or the Australian are similar in nature, with the same potential payoff for participating in planetary colonization.

This list of so-called "syndromes" cannot of course be anywhere near definitive or complete. Some new motivation born perhaps of the entirely new psychological impacts of the Space Age may come to dominate our great-grandchildren's inspirations. Some analogy from non-Western history (perhaps from accounts of activities in Polynesia, Arabia, Sinkiang, or Siberia) may prove more fruitful. There is, after all, no lack of models. There are many lessons yet to be realized from our own past with relevance towards the future, and the part that terraforming will play in that future.

Summing up these analogies with macro-engineering projects and with different modes of colonization, we see some patterns emerging. The energies of the "misfits" and the dreamers are the source of most of the world's major advances; despite the fact that some of these advances have been of questionable value and of great human cost, such activities have been the vanguard of our civilization. While some would question the reliance on these basic human drives whose origins we cannot understand, those drives have been behind every other such advance—and there is no reason to expect them to fail now. It *is* entirely plausible that some future Earth culture will decide to undertake such a massive project as the terraforming of Mars or Venus, and that a large fraction of the culture's population will choose to risk their lives in the project's accomplishment. It would be far more astonishing, looking at the record of human history, if nobody *ever* wanted to terraform a planet. Human schedules are unguessable; the planets can wait.

The Practical Objection

In May 1979 the United Press International news wire carried a story by Bruce Nicholls of the Houston bureau, detailing the terraforming colloquium and quoting its participants. On May 23, the *Houston Chronicle* published an editorial entitled "Why Can't We Just Start with Earth?" which was evidently written in all seriousness (*not* as a parody). After reviewing the basics of the proposals for rebuilding Mars, the editorial continued. "This, to us, is thinking just about as big as it is possible to think and we have nothing but admiration for the imagination of this man. But then we pick up the paper and read about the crises that attend our current fuel shortage and we are sorely tempted to ask the same question we hear others ask: 'Why is it, if we can get men on the moon, and if we have engineers who can conceive of terraforming a planet, why can't we come up with a brand of technology that will solve the fuel problem that plagues us here on Earth?' "

Here again, questions are raised for which no answers are really expected. The basic confusion between the technological and socio-economic problems, and between the completely different ways of attacking them and of judging success in the solution, remains a potential political roadblock for making terraforming happen—especially if world opinion-makers remain as confused as the editors of the *Houston Chronicle*. I would like to remind those editors that new technological challenges have the characteristic of giving birth to new technological capabilities, which meet the challenge and then spill over to enrich the entire civilization wise enough to grapple with such challenges. That, too, is a little-appreciated lesson of the past.

The Space Colony Objection

In the surge of recent interest in space colonies, planetary colonization has received short shrift. Any planetary surface, say proponents of giant rotating structures in free space, is a lousy place for a civilization to be.

According to Professor Gerard O'Neill of Princeton University, "the prospects for colonization of other planetary surfaces are unappealing."[3] O'Neill, who independently re-invented the concept of giant spinning cities in space described earlier by Tsiolkovskiy and Cole, listed the drawbacks of planetary surfaces: their small total areas (only a few times as large, in total, as all of Earth's surface), their attenuated and daily eclipsed sunlight, their full-time gravity, and the energy costs associated with travel both on the planet and away from it.

Dr. Thomas Heppenheimer, another space colonization advocate, discussed terraforming briefly in his book *Colonies in Space* (1976). According to Heppenheimer, the theme grew out of science fiction literature, but "it turned out there were several things wrong with this [idea]. . . ." For example, "the idea of using a planet to provide gravity and to hold an atmosphere really represents the hard way to go about doing these things. Really tremendous amounts of material must be collected, enough to make a planet 5,000 miles [8,000 kilometers] in diameter, before there is enough gravity to hold down an atmosphere and keep it from leaking into space. . . ."

As we have seen, Heppenheimer's estimate of the planetary mass needed to hold down an atmosphere is too large by at least a factor of ten. But he put his finger on a potential technological breakthrough which could make terraforming a far less appealing option for the future: the discovery or invention of a workable alternative system to create land surfaces with gravity, with breathable air, and with protection from space radiation.

Someday the very fabric of the universe may be manipulated by unimaginable machines and energies, so that gravitational forces can be turned on and off at will. Someday small asteroids may be enveloped in molecule-thick protective shells to hold air under pressure (as in Gregory Benford's scheme for the Jovian satellites). Someday other exotic technological developments may offer alternatives to planets as comfortable habitats. But according to space colonization enthusiasts, they already have the answer.

For Heppenheimer and O'Neill and others of the space colony persuasion, the obvious answer is to have large (tens of kilometers in length, and kilometers in diameter) structures with their air and land surfaces on the inside, their pseudo-gravity provided by spinning the structures, and their air held in by pressure bulkheads. Such a development does seem technologically feasible to me, even if its economics are no more clear than terraforming's. Space colonies should not for a moment be considered as a competing concept vis-a-vis terraforming, although over enthusiastic advocates of either may all too readily denigrate the other.

Although these two concepts of space migration are potentially complementary, there remain numerous stumbling blocks ahead of both. Since terraforming has already been honestly catalogued, a few words about space colonies is called for. First, the criticisms leveled against planetary colonization are not serious.

The inhabitants of a terraformed planet would not be reliving Earth's industrial revolution, or creating a medieval subsistence economy. They would be supported by advanced industries in exactly the location O'Neill pinpoints as optimum: deep space. So the energy problems and transportation problems inherent in living on a planet may not be significant. Nor would the 'limited' surface area of terraformed planets be a drawback: terraformed worlds would *not* serve as a dumping ground for Earth's surplus population (neither could space colonies), since Earth must master

254   New Earths

The capacity to rebuild planets also entails the ability to save planets from natural cataclysms, or to actually create them.

the dangers of overpopulation within this century without recourse to any far-off relief valve offered by the future. Those population pressures will already have been solved—or will already have destroyed our civilization—by the time terraforming becomes practical.

O'Neill-type colonies have their own potential problems, not the least of which is the psychological one of whether or not such closed environments would be psychologically acceptable indefinitely to the colonists. Also, the prospect of building stable small-scale closed ecologies is clouded, since small systems do not have the inertia of planetwide systems: inertia which could protect the system from outside perturbations. Space colonies are also exposed to solar and cosmic radiations at levels which could prove difficult to handle.

Is Terraforming Dangerous?

There is one observation about planetary engineering which must be faced squarely (it was mentioned obliquely in the description of the "Unemployed Soldier Syndrome" justification for planetary colonization): the ability to rebuild planets can be applied to any planet, including

Earth, and can be used for any purposes, including destructive ones. The possibility that new scientific and engineering breakthroughs can also have military applications is, sadly, not new; it has often been raised as a justification for the suppression of any scientific or engineering progress. For terraforming, it will be a key issue.

It should not be hard to imagine how the giant machines of planetary engineering can be turned into weapons. Indeed, the mirrors which may ultimately control hurricanes, illuminate rescues, save crops, and ultimately warm distant planets were first conceived of as a weapon of war for burning enemy countries. Any discussion of moving asteroids close to Earth, or of impacting them on other planets, must have disclosed that impacting them on Earth could be a terrible act of war (Dandridge Cole gave this warning years ago and was strongly criticized for allegedly inventing a new doomsday weapon when all he had wanted to do was alert the world to the possibility). Manipulations of planetary rainfall patterns are another obvious weapon, on a much larger scale than the American attempts to induce torrential rainfall over the North Vietnamese invasion route through Laos during United States' involvement in one of the recent wars. Manipulation of the ozone layer and other components of the stratosphere could, for example, pour damaging ultraviolet radiation onto an enemy country, continent, or planet.

Making sterile planets habitable, then, is just another side of the coin to making habitable planets (such as Earth) sterile. These possibilities are chilling, but the study of terraforming will alleviate, not accentuate, such a threat. Already, there are numerous ways to induce planet-wide catastrophes, and the topic of terraforming cannot be blamed for them. Instead, proper study of potential planetary engineering techniques can help recognize attempts at eco-sabotage before they get too far, and can provide countermeasures to repair the damage. Hence the study of terraforming becomes urgent from yet another motivation.

Ethical Aspects of Terraforming

Concerns about potential objections to terraforming from conservationists and environmentalists are apparently well founded. Speaking to the terraforming colloquium in Houston, Dr. Jeffrey Warner of NASA's planetary science office in Washington opened his talk on "Ethical Aspects of Terraforming" with a candid warning. "When I first became aware of this topic I had the environmentalist knee-jerk reaction: the concept was morally wrong and even discussing it would be a sin. Humanity has already screwed up Earth. What right does humanity now have to screw up other planets, especially elegant ones like Mars, Venus, Io, and Titan?"

His opening attention-getting broadside was only partially tongue-in-cheek, Warner revealed, but his subsequent remarks indicated a gradual change in sentiments. "As I started giving terraforming serious consideration, I came to the conviction that terraforming was definitely not categorically wrong, and might even be correct in many cases. . . ."

Weaving a pattern of human destiny based on past analogies, Warner examined the potential effects of terraforming activities on the Solar System, on Earth, and on the human race. His conclusion was in the affirmative, that human expansion into the Solar System was deeply rooted in tradition; furthermore, such expansion (of which terraforming was to be an integral part) would be part of the next step in evolution. "Once Homo Sapiens has colonized other planets, that will surely be a trigger for the episodic acceleration in evolution to Homo Sapiens Prime. I see a grand future for our progeny."[5]

Such optimism, even evangelism, cannot be expected from all enthusiasts of the environmental movement. An opposing viewpoint extracted from an article by biologist Dr. David Thompson with the provocative title "Astropollution" (*CoEvolution Quarterly,* summer 1978) follows.

> The groundwork for conservation of resources of the solar system, including space and celestial bodies, must be laid *now,* before exploitation begins and vested economic interests develop.
>
> [The ideological basis for Thompson's urgent stance was not disguised] Overzealous pursuit of an economic frontier in space will undermine the development of zero-growth philosophies and economies on Earth, spark new demands for Earth's resources, and create additional sources of environmental pollution. . . . If the public is led to believe today that limitless resources of outer space are there to bail us out when the going really gets tough, then we may never be able to achieve zero population and economic growth, which I believe is essential *whether or not* we develop the resources of outer space.

Thompson's article claimed to define the new science of "astroecology," and sported such subheadings as "Conservation of 'Lifeless' Environments" and "Goals for Conservation of the Solar System." Regarding the dismissal of environmentalist objections because the other planets are dead ("there are no snail darters on Mars"), Thompson objects nonetheless.

> There are many resources which need protection and wise management besides the biological resources. These resources of the Solar System [include] raw materials, beauty, and records of the past. . . . Even if the Moon or other planets are truly 'dead,' the virgin state of the environment should still be preserved in places for science. [Later, he addressed the issue of terraforming head-on.] There are important philosophical questions about the conservation of other celestial bodies. Is it right to engage in planetary engineering—to alter the environments of whole planets to suit our needs? Or, would these better be left as international parks?
>
> [Without answering that question (or is there any doubt about the answer Thompson expected?), the world's first "astroecologist" outlined a plan of action to preserve the Solar System against industrial rampages.] Although the Space Treaty of 1967 and later U.N. agreements and conventions prohibit contamination by microorganisms, obligate a nation to pay for damages caused by its satellites, and ban atomic weapons deployment and testing in space, they are only a beginning. Additional international agreements with 'teeth' are needed to ban development of military hardware in space, prohibit nuclear rockets and disposal of nuclear wastes in space, . . . control industrial development, mining, and tourism, and set aside specially protected areas.[6]

Thus the philosophic groundwork has been laid for a painful confrontation between environmentalists and would-be terraformers. It need not be so if advocates concentrate on common goals instead of cross-purposes and realize the essential similarities of many parts of their value systems. We have several generations to sort this out, but it's not too early to start now, before positions solidify and dogmas are born.

The strongest common thread—which might be strong enough to bind an alliance rather than divide enemies—is a mutual desire by both environmentalists and terraformers to see more opportunities for people to choose to live close to "natural" conditions ("natural" in terms of *Earth* conditions), as opposed to the urban anthills already so much in evidence in our century. The astro-ecological concept of parks and forest management entails the notion of a forest ranger who harmoniously orchestrates the flow of natural cycles through the area in question; terraforming would in this sense be "rangering" on a planetary scale. Furthermore, the kinds of people who would want to go to a new terraformed world would probably *not* have any desire to duplicate the urbanized society from which they had fled. In the best of all possible futures, wild areas on Earth would be maintained and expanded; in a more realistic future, dwindling wild areas on Earth may have to be replaced with wild areas created on other planets; in a practical world, the pressures on the reduction of wild areas on Earth may be *diverted* to other planets instead.

New Names for New Earths

It is a common practice to signify a major change in a person's life by adopting a new name. Saul became Paul, Abram became Abraham, Ulyanov became Lenin, Dzhugashvili became Stalin, Wotyla became John Paul II. The same symbolic transformation applies to geographic names as well: Britain became England became Great Britain; Constantinople (originally Byzantium) became Istanbul; Persia became Iran; New Amsterdam became New York.

We are now, thanks to astronomical studies and space probes, quite familiar with the characteristics of the Moon, Mars, Venus, and other worlds. We know their weather, their movements, their suitability (if any) for life.

No greater transformation could be imagined for these planets than to be terraformed. No greater personal or geographic conversion has ever been seen in human history. The planets would be completely changed, and all the characteristics associated with the previous names would be gone or modified.

Signifying such a complete break with the sterile past, new names for these new Earths might be appropriate. Some continuity might be preserved via the thread of mythology. Mars could become Ares, Venus could become (and truly deserve the name) Aphrodite. The Moon might be better known as Selene once it has air to be breathed by native human beings who stroll at night in the earthlight.

Summing Up

We have first dodged the question of "why?" by making the excuse that giant engineering projects of the past have usually been motivated by irrationalities particular to each civilization, so any description of possible motivations a century or two from now would be hopelessly speculative. However, we then did offer half a dozen syndromes from past history which could in some combination serve as analogies for terraforming and colonization. There have been more than enough such analogies to lead us to suspect that there *will* be adequate reasons found, when the time comes to make the decisions.

As to "who?", the same reasoning applies. We have many analogies from the past of the kinds of people who would want to go to a new frontier. They are not always the kinds of people best suited for the frontier, but they pay the consequences. They are the kind of people who for various reasons wish to foresake the parent society—and which the parent society is generally happy to be rid of. They are the cutting edge of human civilization, and the frontiers which they settle have impacts on the mother society out of all proportion to the numbers of people involved and to the amount of resources expended. That has been the lesson of our history: societies that have engaged in colonization have flourished, while those that have not (particularly Germany, Japan, and Italy) have turned these energies to militarism or back in upon themselves with often tragic results. Which fate awaits Earth, only the future will record.

Now, what right does humanity have to take possession of other worlds? What right? To some, the notion of Earth life expanding into the Solar System is an almost mystic crusade; to others, it is a symptom of a virulant cancer growing beyond its normal bounds. I have put this question last not because my suggested answer is the most profound part of the book, but because few people in history ever get to this question. Human history demonstrates that people do what suits them and seek rationalizations afterwards. There is no reason to suspect that terraforming will be any different, or that the lack of a good answer to this question will have the slightest impact on what people choose to do on other planets. If terraforming becomes part of our future, justifications will be found.

> Upon a slight conjecture I have ventured on a dangerous journey, and already I behold the foothills of new lands. Those who have the courage to continue the search will set foot upon them.
>
> Immanuel Kant, 1754

Afterword

Terraforming is possible. Given the appropriate energy, resources, and time, many of the now-sterile worlds of the Solar System can be rebuilt into new Earths.

This is not to deny the tremendous technological problems facing such planetary engineering projects. As we have discovered on our dangerous journey, there are urgent questions in planetary geology; in mineralogy and soil chemistry; in agriculture and ecology; in biology and genetics; in atmospheric physics and meteorology and climatology—questions which we have only begun to ask. Doubtless there are numerous other problems still unrecognized.

But lined up against such stumbling blocks is human ingenuity. Any book written towards the close of the twentieth century cannot presume to forecast the actual arsenal of human capabilities available in the twenty-second or twenty-third centuries, or even just fifty years from now. We have, however, been able to assemble a collection of ideas and schemes which should have demonstrated that for every roadblock there is a detour, for every wall there is a trap door or a ladder or a big enough bomb.

This book has not been a blanket advocacy of planetary engineering, but an advocacy of the investigation of the possibilities for planetary engineering. Questions raised in these intellectual, scientific, and technological pursuits will have applications elsewhere in our civilization. Detailed research into planetary engineering, into terraforming, even if it never occurs in any form remotely resembling the speculations put forth in this book, will still be useful.

Possible motivations for such activities remain obscure, and understandably so. Who could have forecast the social pressures which built pyramids, cathedrals, and moon rockets? Some may fear the irrationality which may make terraforming a part of our descendants' civilization; others may value it as a human trait in an era of all too impersonal technology.

Simply as an exercise in speculation, and as a target for criticism and for the better ideas which lie dormant in the minds of thousands of readers, I have drawn up a possible schedule for planetary engineering over the next few centuries and revealed it in the fictitious vignettes in this book. To be provocative, I've been as optimistic as I dare, while still being even marginally credible. But I remain fully aware of the dangers of falling short of reality, and of being a victim of a failure of nerve, and of not counting on the unexpected—a failure which has made practically all predictions of future life written a century ago to appear today as hopelessly shortsighted and unimaginative. Risking all that, I trust that readers a century from now will forgive my all-too-limited imagination for the sake of my unlimited faith.

Notes

Introduction
1. James E. Oberg, ed., *The Terraforming Papers*. (Forthcoming)
2. Fritz Zwicky, *Morphological Astronomy*, p. 37.
3. Fritz Zwicky, "The March into Inner and Outer Space," *Engineering and Science* June (1961).
4. Carl Sagan, *The Planet Venus*, p. 233.
5. Freeman J. Dyson, *The Search for Extrasolar Technology*, p. 10.
6. M.M. Averner and R.D. MacElroy, *On the Habitability of Mars—An Approach to Planetary Ecosynthesis*, p. 14.

Chapter 1
1. Isaac Asimov, *A Choice of Catastrophes*.

Chapter 2
1. Stephen H. Dole and Isaac Asimov, *Planets for Man*, p. 76.

2. J.E. Lovelock, *Gaia: A New Look at Life on Earth*, p. 84.
3. Owen B. Toon and James B. Pollock, "Atmospheric Aerosols and Climate," *American Scientist* May–June (1980): pp. 268–77.
4. Reid A. Bryson, "Volcanic Activity and Climatic Changes," p. 5.
5. Henry Stommel and Elizabeth Stommel, "The Year Without a Summer," *Scientific American* June (1979): pp. 176–86.
6. D.C. Hallacy, Jr., *Ice or Fire: Surviving Climatic Change*, p. 126.
7. Charles R. Carrigan and David Gubbins, "The Source of the Earth's Magnetic Field," p. 126.
8. Alan Hammond, "The Uniqueness of Earth's Climate," p. 126.

Chapter 3

1. Scientific American, *The Biosphere*.
2. Fred Hoyle, *Astronomy and Cosmology*, (San Francisco: W.H. Freeman 1975): p. 196.
3. Godfrey T. Sill and Laurel L. Wilkening, "Ice Clathrate as a Possible Source of the Atmosphere of the Terrestrial Planets," *Icarus*, vol. 33 (1978): pp. 13–22.
4. James B. Pollack and Y.L. Yung, "Origin and Evolution of Planetary Atmospheres," *Annual Review of Earth and Planetary Science* (1980): pp. 425–87.
5. Michael Hart, *The Evolution of the Atmosphere on Earth*, p. 30.
6. J.E. Lovelock, *Gaia: A New Look at Life on Earth*, p. 40.
7. Alan Hammond, "The Uniqueness of Earth's Climate," p. 245.
8. Richard A. Kerr, "Climate Control: How Large a Role for Orbital Variations?" p. 144.
9. Study of Man's Impact on Climate Committee (1971) p. 56.
10. Stephen H. Schneider with Lynne E. Mesirow, *The Genesis Strategy: Climate and Global Survival*, p. 228.

Chapter 4

1. Stephen H. Schneider with Lynne E. Mesirow, *The Genesis Strategy: Climate and Global Survival*, p. 206.
2. John Gribbin, "Woodman, Spare That Tree," *New Scientist*, Mar 29 (1979): pp. 1016–8.
3. *Science News*, (1969) (96: 330).
4. *New Scientist*, Apr 19 (1979): pp. 196–199.
5. Richard Brook Cathcart, *The Developing Artificial Geography of the Solar System*.
6. Willy Ley, *Engineer's Dreams*, p. 177.

7. Mark L'vovich, "Turning the Siberian Waters South," p. 834.
8. D.S. Hallacy, Jr., *Ice or Fire: Surviving Climatic Change,* p. 129.
9. E.C. Barrett, *Climatology from Satellites,* p. 144.
10. Gordon R. Taylor, *The Doomsday Book,* p. 30.

Chapter 5

1. B. Donn, et al., *The Study of Comets,* NASA SP-393 (Washington, D.C.: US Government Printing Office 1976).
2. Dandridge Cole, *Islands in Space,* (Philadelphia: Chilton Press 1964): p. 36.
3. Brian O'Leary, *The Fertile Stars.*
4. David Morrison and William C. Wills, eds., *Asteroids: An Exploration Assessment.*
5. Ibid pp. 160–4.
6. Ibid pp. 148–54.
7. A.R. Martin, *Project Daedalus: The Final Report on the B.I.S. Starship Study,* pp. 66–7.
8. M.M. Averner, and R.D. MacElroy, *On the Habitability of Mars—An Approach to Planetary Ecosynthesis,* pp. 48–59.
9. Carl Sagan, *The Planet Venus,* p. 233.

Chapter 6

1. K. Eric Drexler, "Gossamer Spacecraft."
2. Brian O'Leary, *The Fertile Stars.*
3. Louis A. Kleiman, *Project Icarus.*
4. Recent development of the idea is attributed to Jerome Pearson of Wright Patterson AFB, Charles Sheffield, and Hans Moravec, "Skyhook!," pp. 1–3.
5. E.C. Okress, *Construction Specifier.*
6. *Future Space Programs* 1975, Stock Number 052-070-02889-1 (Washington, D.C.: US Government Printing Office 1975): pp. 349–64.
7. Chorley and Moore are quoted in Gordon R. Taylor's *The Doomsday Book,* p. 28.
8. "Earth May Have Had Saturn-Like Rings 34 Million Years Ago," NASA press release 80–86 June 9 (1980).
9. "Sun Earth Explorer Near Libration Point," *Aviation Week,* Aug 21 (1978) p. 17.

Chapter 7

1. John Gribbin, "Martian Climate: Past, Present, and Future," pp. 6–11.

2. James B. Pollack, "Climatic Change on the Terrestrial Planets," pp. 479–553.
3. Owen B. Toon, et al., "Climatic Change on Mars: Hot Poles at High Obliquity," *American Astronomical Society Bulletin*, vol. 9 (1977): p. 450.
4. Carleton C. Allen in Ronald Greeley, ed., "Second Mars Colloquium," pp. 468–472.
5. Thomas A. Mutch, *The Geology of Mars*.
6. Joel S. Levine, "A New Estimate of Volatile Outgassing of Mars," *Icarus*, vol. 28 June (1976): pp. 165–9.
7. Joseph A. Burns and Martin Harwitt, "Towards a More Habitable Mars—or—The Coming Martian Spring," p. 127.
8. M.M. Averner and R.D. MacElroy, *On the Habitability of Mars—An Approach to Planetary Ecosynthesis*, p. 14.
9. Kunze—personal communication

Chapter 8

1. Arthur C. Clarke, *The Fountains of Paradise*, (New York: Harcourt Brace Jovanovich 1979): pp. 116–7.
2. A. Kliore, ed., *The Mars Reference Atmosphere*, pp. 146–8.
3. R.D. MacElroy, "A Post-Viking Reassessment: The Habitability of Mars," pp. 6–7.
4. C.W. Snyder in Ronald Greeley, ed., "Second Mars Colloquium."
5. R.D. MacElroy, "Post-Viking Reassessment," p. 8.
6. Ibid.
7. Christopher P. McKay and Steven M. Welch, "The Response of the Martian Climate to Natural and Artificial Perturbations," p. 6.
8. Strickland, personal communication
9. M.M. Averner and R.D. MacElroy, *On the Habitability of Mars—An Approach to Planetary Ecosyntheses*, pp. 81–3.
10. James E. Oberg, ed., *The Terraforming Papers*. (Forthcoming)
11. Stephen H. Dole and Isaac Asimov, *Planets for Man*, p. 36.
12. Stephen H. Dole, personal communication
13. Mars topographic map #I-961 (Flagstaff, AZ.: U.S. Geological Service, 1976).
14. Hans R. Thierstein and Wolfgang H. Berger, "Injection Events in Ocean History," pp. 461–6.
15. Ben Bova, ed., *Closeup: New Worlds*, p. 82.

Chapter 9

1. James E. Oberg, "Venus," *Astronomy*, Aug (1976): p. 9.

2. Peter Pettengill, et. al., "The Surface of Venus," Scientific American, Aug (1980): pp. 54–65.
3. James B. Pollack, "Climatic Change on the Terrestrial Planets," Icarus, vol. 37 (1979): p. 526.
4. Carl Sagan, "Planets," pp. 240–2.
5. Jerry Pournelle, "Step Further Out—The Big Rain," Galaxy, Oct (1974): pp. 54–60.
6. Steven M. Welch and Christopher P. McKay, "Venus: The Terraforming View." Paper read at the Colloquium on Terraforming, March 17, 1979 at Houston, Texas.

Chapter 10

1. Dandridge M. Cole, Beyond Tomorrow, p. 111.
2. Vondrak, personal communication.
3. Richard R. Vondrak, "Creation of an Artificial Atmosphere on the Moon," p. 177.
4. John O'Keefe, "Tektites," Scientific American, Aug (1978): pp. 116–125.
5. Hollenbeck, personal communication.
6. P. Cassen, et al., "Melting of Io by Tidal Dissipation," Science, vol. 203 (1979): pp. 892–894.

Chapter 11

1. Frank P. Davidson, et al., "Macro-Engineering and the Infrastructure of Tomorrow," p. 26.
2. Paul Meadows, attributed this observation to a study by Dr. Rosabeth Kantor in Community in Space.
3. Gerard O'Neill, The High Frontier, p. 240.
4. Thomas Heppenheimer, Toward Distant Suns, pp. 44–5.
5. Jeffrey L. Warner, "Ethical Aspects of Terraforming," pp. 1–4.
6. David Thompson. "Astropollution," CoEvolution Quarterly, Summer (1978): pp. 35–48.

Glossary

ADVECTION—The horizontal transport of heat across the face of a planet by atmospheric winds.

AEOLIAN—Referring to processes carried out by wind, such as erosion or dust deposition.

AEROSOL—Liquid material suspended as droplets in the atmosphere.

AEROSTAT—Large, high-altitude free-floating structures filled with a constant-volume of hot air capable of supporting very heavy payloads.

ALBEDO—The reflectiveness of a planet's surface, from zero (full absorption, or black) to 100% (full reflection, or pure white) of incoming sunlight, which is rejected back into space.

ASTROBLEME—"Fossil crater"; actually the geologic traces of the meteor or asteroid impact which can be detected after hundreds of millions of years of Earth-type erosion and tectonism.

ASTRONOMICAL UNIT—Classical relative yardstick of the Solar System (before space probes and interplanetary radar produced highly accurate measurements in absolute mileage), equivalent to the average Earth-Sun distance, equal to about 93,100,000 miles or 8 light-minutes.

ATMOSPHERE—The collection of gaseous material surrounding a planet.

BAR—Measurement of atmospheric pressure, originally set in units of Earth sea level pressure (about 15 pounds per square inch).

BIOMASS—Organic material, both living and dead.

BIOSPHERE—Region of a planet where life is active; more hostile regions where life can survive but is inactive compose the "parabiosphere."

BIOTA—Living organisms.

BURNS-HARWITT MANEUVER—The process of emplacing large satellites around a planet to stabilize the precession of the planet's axis.

CARBONACEOUS—Material rich in the volatile elements such as carbon, nitrogen, oxygen, and hydrogen.

CELESTIAL MECHANICS—The science of the mutual movements of celestial objects such as planets, asteroids, and spacecraft under the influence of gravity.

CLIMATE—The time-averaged weather conditions of a region.

CONVECTION—The vertical transport of heat within a planetary atmosphere by means of upward and downward moving air masses.

CORIOLIS EFFECT—The tendency of north-south movements to appear to be deflected to the east or west due to the rotation of a planet; explains the patterns of wind belts and cyclonic storms.

CRYOSPHERE—The portion of a planet dominated by frozen water.

CYANOPHYTES—Blue-green algae, a plant particularly well suited to life in adverse, unearthly environments.

DELTA-VEE—The velocity change for an object being navigated through space, to alter its course.

DIFFERENTIATION—Process by which a planetary body of intermixed materials becomes separated into materials or different weights, forming a denser core and a lighter crust.

ECCENTRICITY—The degree of departure of an orbit from a perfect circle. $E=0$ for a circle, $E=1$ for a parabolic (open) orbit, $E>1$ for a hyperbolic (or very open) orbit.

ECOLOGY—The science of the interrelationships of factors in a biosphere; or the actual pattern of all life forms in an environment.

ECOSYNTHESIS—The artificial creation of an ecology through the application of energy, matter, and ingenuity.

EON—Geologic shorthand for a period of one billion years.

Glossary

ESCAPE VELOCITY—Velocity needed for a projectile to leave the sphere of gravitational dominance of an object, on a parabolic path.

EVAPOTRANSPIRATION—Process by which plants, and in particular trees, draw water up through their roots, then release it into the air as vapor.

FEEDBACK LOOP—A process which is controlled by factors which themselves are affected by the process itself; a loop can either have positive feedback, in which small deviations cause increasingly large ones, or negative feedback, in which small deviations create restoring, balancing forces.

FLYWHEEL—A mechanical system designed to resist changes in motion and thus store momentum; in terraforming, a system with high inherent stability requiring long and powerful perturbation in order to change it.

FULCRUM—In mechanics, the point about which a lever turns in moving an object; metaphorically in terraforming, a unique aspect of an ecosphere at which relatively small actions can have major climatic consequences.

GAIA HYPOTHESIS—The suggestion that the biosphere of a planet acts (via variations in the "greenhouse effect," among other ways) to modulate planetary surface conditions to keep them conducive to life.

GALACTIC ENGINEERING—Wholesale rebuilding of galaxies, involving energy releases, star formation in mass, alteration of the mass distribution and energy flux of large portions of the galaxy, and other processes which would be easily visible millions of light-years away.

GARDENING—Turning over a planet's regolith to expose fresh subsurface material to atmospheric weathering, erosion, and chemical interaction.

GEOSYNCHRONOUS—An orbit around Earth with a revolution period equal to the time Earth rotates once; above the equator, a geosynchronous satellite would appear to hang in the same place in the sky. Synchronous satellites are possible over planets with short "days," such as Mars or Jupiter or Saturn, but not over Venus, Mercury, or the Moon because their slow rotation rates would require such a high (and hence slow) satellite orbit that the Sun's gravity would cause the satellite to escape.

GRAVITY WELL—Metaphorical term for the energy hole out of which a spaceship must be propelled in order to escape the planet and drift free in interplanetary space.

GREENHOUSE EFFECT—Metaphorical term for the very real elevating effect of a planetary atmosphere on surface temperature; the level of elevation depends on the combination of various gases and aerosols.

HEAT TRAP—Any system which absorbs and stores large quantities of thermal energy.

HYDROSPHERE—That portion of a planet's surface consisting of liquid water.

INSOLATION—Amount of solar energy arriving at a planet; it is modulated by the planet's albedo.

ISOSTATIC—Evenly balanced between geologic forces such as a granite mountain range pushed up by crustal motion and buoyancy, and pulled down by gravity.

LAMINAE—Layers of alternating material laid down in the Martian polar regions by some unknown mechanism at some unknown date in the distant (?) past.

LOESS—Fine-grained wind-blown dust forming thick deposits, notably in northern China; it is both evidence of major past climatic variations and, while suspended in the atmosphere, was probably a contributing factor to such climate changes.

MACRO-ENGINEERING—Creation of giant projects lasting for decades and costing appreciable fractions of a society's gross national product; in general, they are not practical but serve some social, religious, psychological, or other purpose.

MAGMA—Underground molten rock.

MILLIBAR—Unit of atmospheric pressure equal to .001 of a bar.

MODEL, COMPUTER—A formal description of the factors which contribute to a system, including their initial conditions and their interrelationships, programmed into a computer so as to allow prediction of future trends, possible variations in actual initial conditions or relationships, and relative importance of different factors in the model.

MONTMORILLONITE—A hydrous aluminum silicate, named after the French village where it was first found and classified.

OBLIQUITY—Angle of deviation between a direction of interest (say, the pole of a planet's rotation) and a standard direction (say, the line perpendicular to the planet's orbital path around the sun).

ORBITAL VELOCITY—Velocity horizontal to the surface of a planet, at which an object's fall towards the surface under the influence of gravity is equal to the rate of retreat of the surface due to the roundness of the planet (such an object is in "free fall," or is "weightless"—it is *not* undergoing "zero gravity"). Velocity equals at least 71% of escape velocity.

OUTGASSING—Process of release of gas or vapor from a solid or liquid material, accelerated by heating or drop in pressure, usually through vents or volcanoes.

OXIDATION—The combining of material with oxygen (e.g. rusting or combustion).

OZONE—A form of oxygen molecule containing three oxygen atoms linked together (normal atmospheric oxygen consists of pairs of atoms) which absorbs solar ultraviolet radiation, protecting surface biota.

Glossary

PANSPERMIA—Theory that Earth life is only a local manifestation of an ecology which is spreading (naturally or artificially) through the whole galaxy.

PERMAFROST—Layer of permanently frozen soil beneath a planet's surface, extending from below the near-surface zone which thaws each summer down to the depths warmed by the interior heat of the planet. May contain vast reservoirs of ice intermixed with soil.

PERTURBATION—Disturbance by an outside force or action on a closed system, such as an ecosphere or a trajectory.

PHOTOSYNTHESIS—Process by which green plants use solar energy to convert inorganic material (including water and carbon dioxide) into biomass, giving off oxygen as a byproduct.

PLANETARY ENGINEERING—Artificial activities on a planet-wide scale with planet-wide effects, such as terraforming.

PLANETESIMALS—Small bodies in the early Solar System which were the building blocks of the large planets.

PRECESSION—Gradual rotation of the plane of a planet's orbit, or the axis of the planet's rotation, caused by gravitational effects of nearby objects on the planet or on bulges within it.

PROGRADE—Orbital motion around an object in the same direction as the majority of the rest of the satellites (when viewed from north, counterclockwise). Also called "posigrade;" opposite of "retrograde."

PSEUDOGRAVITY—Acceleration forces within spinning space vehicles which mimic gross effects of planetary gravity.

QUARANTINE—Biological isolation of a sample from a planet (such as the Apollo moon rocks), or of a planet from a manmade object (such as the prevention of biological contamination of Mars by space probes); the philosophical opposite of terraforming.

REFRACTORY—Elements with high melting and boiling points, the kinds most likely to condense first close in to the hot, young sun; the opposite of volatile.

REGOLITH—The layer of debris which covers the face of a planet, if inorganic; if capable of supporting biota, such a layer deserves the name "soil."

RETROGRADE—An orbital path around a primary object which is opposite in direction to the orbits of its regular satellites. Viewed from north, it would be clockwise.

SETI—The Search for Extraterrestrial Intelligence (as opposed to the more active Communications with ETI, or CETI—pronounced the same), usually reserved for attempts to detect alien high-technology activities such as radio beacons, lasers, and thermonculear warfare.

SINK—A region or mechanism for the absorption of a liquid or a gas.

Glossary

SKYHOOK—Fanciful space device consisting of a base station in orbit and a superstrong cable let down to or near the surface. Rotating skyhooks, described by Moravec, do not require a base station in geosynchronous orbit. Also called "space elevator" (Clarke) or "beanstalk" (Sheffield). The required strength of materials—several orders of magnitude higher than currently available—make Earth-side skyhooks impossible at present; however, rotating skyhooks over smaller bodies such as the Moon or Mars are conceivably within the reach of the early twenty-first century.

SWINGBY—Space trajectory maneuver involving close pass to a massive planet in order to effect gross changes in the orbital path and energy of a guided space object.

SYNERGY—The effect by which a combination of forces or actions has a much greater net effect than merely the sum total of the effects of each force or action applied separately.

TECTONISM—Process of "continental drift" in which plates drift across the face of planet, giving rise to mountain ranges, rift valleys, mid-basin ridges, and deep trenches.

TERRAFORM—Artificially perturb the energy material inventory of a planet in order to create or preserve conditions capable of supporting terrestrial life forms including human beings.

TROPOSPHERE—Atmospheric layer within which temperature decreases with increasing altitude; at a point called the tropopause, this trend reverses, thus creating a trap which helps Earth hold onto its water vapor.

ULTRAVIOLET—Spectrum of radiation beyond the blue end of the visible light band, extremely damaging to life forms.

VITRIFICATION—Conversion of a material to glass by heat.

VOLATILE—Easily evaporated by heat.

Bibliography

Adelman, S., and Adelman, B. 1981. *Bound for the Stars*. Englewood Cliffs, N.J.: Prentice-Hall.

Arnold, J.R., and Duke, M.B. 1978. *Summer Workshop on Near-Earth Resources*. NASA CP-2031. Washington, D.C.: U.S. Scientific and Technical Information Office.

Asimov, I. 1979. *A Choice of Catastrophes*. New York: Simon and Schuster.

Averner, M.M., and MacElroy, R.D. 1976. *On the Habitability of Mars—An Approach to Planetary Ecosynthesis*. NASA SP-414. Springfield, Va.: National Technical Information Service.

Barrett, E.C. 1974. *Climatology from Satellites*. London: Methuen.

Beaty, C.B. 1978. The Causes of Glaciation. *American Scientist* Jul–Aug: 452–9.

Berry, A. 1974. *The Next Ten Thousand Years*. New York: E.P. Dutton.

Billman, K.W., Gilbreath, W.P., and Bowen, S.W. 1979. Solar Energy Economics Revisited: The Promise and Challenge of Orbiting Reflectors for World Energy Supply. Mountain View, Calif.: NASA Ames Research Center.

Bibliography

Bova, B., ed. 1977. *Closeup: New Worlds.* New York: St. Martin's Press.

Bryson, R.A. 1981. Volcanic Activity and Climatic Changes. Paper read at the Annual Conference of the American Association for the Advancement of Science, 3–8 Jan 1979, at *Houston, Texas.* Virginia Beach: Donning Press.

Burns, J.A., and Harwitt, M. 1973. Towards a More Habitable Mars—or—The Coming Martian Spring. *Icarus* 19: 126–130.

Calder, N. 1978. *Spaceships of the Mind.* New York: Viking Press.

Carrigan, C.R., and Gubbins, D. 1979. The Source of the Earth's Magnetic Field. *Scientific American* Feb: 118–128.

Cathcart, R.B. 1979. *The Developing Artificial Geography of the Solar System.* Monticello, Ill.: Vance Bibliographies.

Cavanaugh, K.M. 1978–81. *The Martian Chronicles: The Newsletter of the Mars Study Project.* Boulder: University of Colorado Press.

Chandler, D.L. 1979. *Life on Mars.* New York: E.P. Dutton.

Cole, D.M. 1965. *Beyond Tomorrow.* Amherst, Wis.: Palmer Publications.

Corn, M.A., McKay, C.P., and Boston, P. 1979. Planetary Engineering of Mars and the Biological Control of Planetary Evolution. Paper read at the Colloquium on Terraforming, March 17, 1979, at Houston, Texas.

Criswell, D.R., ed. 1977. *New Moons: Towing Asteroids into Earth Orbits for Exploration and Exploitation.* Houston: Lunar and Planetary Institute.

Davidson, F.P., Giacoletto, L.J., and Salkeld, R., eds. 1978. Macro-Engineering and the Infrastructure of Tomorrow. In *AAAS Selected Symposia Series.* Boulder: Westview Press.

DeNeni, D.P. 1978. *The Weather Report.* Millbrae, Calif.: Celestial Arts Press.

Dole, S.H. 1964. *Habitable Planets for Man.* Santa Monica: Rand Corporation report R-414-PR.

Dole, S.H., and Asimov, I. 1964. *Planets for Man.* New York: Random House.

Drexler, K.E. 1978. Laser Propulsion to Geosynchronous Orbit. *L-5 News* Jul: 8–10.

———. 1979. High Performance Solar Sail Concept. *L-5 News* May: 5–9.

———. 1980. Gossamer Spacecraft. *L-5 News* Feb: 4–5.

Dyson, F.J. 1965. The Search for Extrasolar Technology. Princeton: Institute for Advanced Study.

———. 1977. The Next Industrial Revolution. *The Key Reporter* Spring: 2–5.

———. 1979a. *Disturbing the Universe.* New York: Harper and Row.

———. 1979b. Pilgrims, Saints, and Spacemen. *L-5 News* May: 5–9.

Ehricke, K.A. 1978a. The Extraterrestrial Imperative. *Space World* Nov: 4–12.

———. 1978b. *A Socio-Economic Evolution of the Lunar Environment and Resources.* La Jolla, Calif.: Space Global Corporation.

———. 1978c. *Space Light: Space Industrial Enhancement of the Solar Option*. La Jolla, Calif.: Space Global Corporation.

Gehrels, T., ed. 1979. *Asteroids*. Tucson: University of Arizona Press.

Golden, F. 1981. Shaping Life in the Lab. *Time* Mar 9: 50–9.

Greeley, R., ed. 1979. Second Mars Colloquium. *Journal of Geophysical Research* Dec 30: 468–472.

Gribbin, J. 1977. Martian Climate: Past, Present, and Future. *Astronomy* Oct: 6–11.

———. 1980. *The Death of the Sun*. New York: Delacorte Press.

Hallacy, D.S., Jr. 1978. *Ice or Fire: Surviving Climatic Change*. New York: Harper and Row.

Hamil, R. 1978. Terraforming the Earth. *Analog* Jul: 47–65.

Hammond, A. 1975. The Uniqueness of Earth's Climate. *Science* Jan 24: 245.

Hart, M. 1978. The Evolution of the Atmosphere of the Earth. *Icarus* 33: 23–39.

Henson, C. 1979. Planetary Chauvinism. *Future Life* Nov: 41.

Heppenheimer, T. 1977. *Colonies in Space*. Harrisburg, Pa.: Stackpole Books.

———. 1979. *Toward Distant Suns*. Harrisburg, Pa.: Stackpole Books.

Kerr, R.A. 1978. Climate Control: How Large a Role for Orbital Variations? *Science* July 14: 144–146.

Kleiman, L.A., ed. 1968. *Project Icarus*. Cambridge: MIT Press.

Kliore, A., ed. 1978. *The Mars Reference Atmosphere*. Pasadena, Calif.: Jet Propulsion Laboratory.

Levine, J. 1977. The Making of an Atmosphere. *Advances in Engineering Science* NASA CP-2001.

Ley, W., rev. ed. 1964. *Engineer's Dreams*. New York: Viking Press.

Lovelock, J.E. 1979. *Gaia: A New Look at Life on Earth*. Oxford: Oxford University Press.

Lunan, D. 1979. New Worlds for Old. New York: William Morrow.

L'vovich, M. 1978. Turning the Siberian Waters South. *New Scientist* Sep 21: 834–836.

MacElroy, R.D., and Averner, M.M. 1977. Atmospheric Engineering of Mars. *Advances in Engineering Science* NASA/#CP-2001.

———. 1980. A Post-Viking Reassessment: The Habitability of Mars. NASA Ames Research Center.

Martin, A.R., ed. 1978. *Project Daedalus: The Final Report on the B.I.S. Starship Study*. London: British Interplanetary Society.

McKay, C.P. 1979. Some Basic Physical and Climatological Considerations in Terraforming Mars. Paper read at the Colloquium on Terraforming, March 17, 1979, at Houston, Texas.

McKay, C., and Welch, S.M. 1979. The Response of the Martian Climate to Natural and Artificial Perturbations. Paper read at the Colloquium on Terraforming, March 17, 1979, at Houston, Texas.

Moravec, H. 1978. Skyhook! *L-5 News* Aug: 1–3.

Morrison, D., and Wills, W.C., eds. 1978. *Asteroids: An Exploration Assessment* NASA CP-2053. Washington, D.C.: U.S. Scientific and Technical Information Office.

Mutch, T.A., Arvidson, R.E., Head, J.W. III, Jones, K.L., Saunders, R.S. 1977. *The Geology of Mars*. Princeton: Princeton University Press.

Oberg, J.E. 1978. Terraforming. *Astronomy* May: 6–13. Reprinted in *A Reader-Study Guide to Carl Sagan's Cosmos*. 1980. New York: Random House.

———. 1979a. Farming the Planets. *OMNI* Feb: 34–37.

———. 1979b. Planetary Engineering and the Search for Extraterrestrial Intelligence. Paper read at the Where Are they? The Implications of Our Failure to Detect Extraterrestrials Conference, 2–3 Nov 1979, at the University of Maryland, College Park, Md. Proceedings to be published by Pergamon Press.

———. Forthcoming. *The Terraforming Papers: Proceedings of the 1979 Houston Colloquium on Terraforming*.

———. Colony on the Planet Epaphos. 1980. *Star and Sky* Mar: 16.

O'Leary, B. 1981. *The Fertile Stars*. New York: Everest House.

O'Neill, G. 1977. *The High Frontier*. New York: William Morrow.

Paltridge, G. 1979. The Problems with Climate Prediction. *New Scientist* Apr 19: 194–195.

Pettengill, P., et. al. 1980. The Surface of Venus. *Scientific American* Aug: 54–65.

Pignolet, G. 1980. Retrieving Asteroids. *L-5 News* Mar: 6–7.

Pollack, J.B. 1979. Climatic Change on the Terrestrial Planets. *Icarus* 37: 479–553.

Pollard, W.G. 1979. The Prevalence of Earthlike Planets. *American Scientist* Nov–Dec: 653–9.

Robock, A. 1978. Internally and Externally Caused Climate Change. *Journal of the American Meteorological Society* June: 436–8.

Sagan, C. 1961. The Planet Venus. *Science* Mar 24: 227–235.

———. 1966. Planets. New York: New York Times, Inc.

Sagan, C., and Shklovskii, I.S. 1964. *Intelligent Life in the Universe*. New York: Holden-Day Press.

Schneider, S.H., with Mesirow, L.E. 1976. *The Genesis Strategy: Climate and Global Survival*. New York: Dell Publishing Co.

Scientific American. 1970. *The Biosphere*. San Francisco: W.H. Freeman and Co. Reprint of the Sep 1970 issue of *Scientific American*.

———. 1977. *Ocean Science*. San Francisco: W.H. Freeman and Co.

Sessions, L. 1980. Preparing Mars for Life. *Science Digest* Nov–Dec: 128–9.

Solomon, S. 1980. Amazing Machines that Move Planets. *Science Digest* Nov–Dec: 56–61.

Study of Man's Impact on Climate Committee. 1971. *Inadvertant Climate Modification*. Cambridge: MIT Press.

Taylor, G.R. 1970. *The Doomsday Book*. Greenwich, Conn.: Fawcett Publications.

Thierstein, H.R., and Berger, W.H. 1978. Injection Events in Ocean History. *Nature* Nov 30: 461–6.

Verschuur, G. 1978. *Cosmic Catastrophes*. New York, Addison-Wesley.

Vondrak, R.R. 1977. Creation of an Artificial Atmosphere on the Moon. *Advances in Engineering Science* NASA CP-2001.

von Puttkamer, J. 1979. In Earth's Image: The Terraforming of Other Planets. *Future Life* Mar: 54–5.

Warner, J.L. 1979. Ethical Aspects of Terraforming. Paper read at the Colloquium on Terraforming, March 17, 1979, at Houston, Texas.

Welch, S.M., and McKay, C.P. 1979. Venus: The Terraforming View. Paper read at the Colloquium on Terraforming, March 17, 1979, at Houston, Texas.

Wolfendale, A. 1979. Cosmic Rays and Ancient Catastrophes. *New Scientist* Aug 21: 316–318.

Zwicky, F. 1948. Morphological Astronomy. *The Observatory* Aug: 32–39.

———. 1960. Some Possible Operations on the Moon. *American Rocket Society Journal* Dec.

Index

advective heating, 89, 164, 176–177
aerostat, 133–135, 218
albedo, 44–46, 58–59, 82–83, 89–90
algae, 25–26, 116–118, 164–166, 168, 186, 206–207
Anderson, Poul, 24, 206
Apollo asteroid, 109
Arctic Ocean ice, 91–93
artificial gravity, 38–39
Asimov, Isaac, 150–151, 189
asteroids
    as excavators, 106, 169–170
    composition, 106–108

earth-crossers, 108–109
hitting Earth, 37
import to Earth, 232
mining, 105
momentum, 106
moving, 129–132
origin, 106–108
atmospheres, 46–50, 71
Averner, Melvin, 173

Barsoom, *See Mars*
bimodal stability, 92
biosphere, 43

280   Index

Boston, Penelope, 186–187, 188–191
Bradbury, Ray, 173
British Interplanetary Society, 115
Burns-Harwit maneuver, 162–163
Burroughs, Edgar Rice, 19

Callisto, 235, 239–240
CIA, 97
Clark, Benton, 179
Clarke, Arthur C., 38, 123, 150, 174–5, 183
climate, See Earth biosphere
Cole, Dandridge, 223, 226, 231, 252
comets, 103–105

Daedalus, 115
dams, 94–95
dinosaurs, 37
Dole, Stephen, 189
Dune, 24, 41, 119–120
dust, 58–60, 79
Dyson, Freeman, 26–28, 40, 242, 248
Dyson sphere, 28

Earth
  acid rain, 88–89
  as target for terraforming, 36–37
  atmosphere, 46–50, 53–60, 71–76
  biosphere, 41–62
  carbon dioxide, 53, 87–88
  carbon monoxide, 88

climate modification, 86
climate variability, 76–80
contaminants, 89
dams, 94–95
human effects on, 80–84
hurricane steering, 98–100
magnetic field, 37–38, 60–62
oceans, 50–51
origin, 63
polar caps, 89–93
rainfall, 51–53, 93
rings around, 141–142
source of volatiles, 65–68
Terraforming the Earth, 29
Egge, David, 34
Ehricke, Krafft, 38, 137–139, 183
Europa, 235, 237–238
evapotranspiration, 84

Farmer in the Sky, 22–23
Ferguson, Eugene, 244–246
fire, 55
Fuller, Buckminster, 135

Gaia Hypothesis, 64, 74–75
galactic engineering, 27, 39
Ganymede, 22–23, 24, 163, 235, 238–239
Genesis, 32, 194–195
genetic engineering, 116–118, 167–168
greenhouse effect, 53–55, 73, 87–88, 166, 205

Handbook of Terraforming, 17
Hamil, Ralph, 29

Hart, Michael, 72–73
Heinlein, Robert, 21, 22–24, 120
Heppenheimer, Thomas, 252–253
Holland, 98
hurricanes, 49, 98–100

Ice Ages, 36–37, 45, 78–80
Io, 235, 236–237

Jupiter, 22, 28, 35, 36, 109, 111–113, 115, 134–135, 235–240

Kunze, A. W. G., 169–170

Lagrange Points, 142
Last and First Men, 20–21
Levine, Joel, 161
Ley, Willy, 94
lichens, 186–187
light sails, 129, 139
Lovelock, James, 74–75
Lowell, Percival, 19

Macroengineering, 244–246
MacElroy, Robert, 181, 185, 187–188
Mars, 25, 149–195
   as terraforming target, 35–36
   atmosphere, 152, 160, 175–176
   Barsoom, 19
   climate history, 153–157
   dust storms, 175–179
   ecosynthesis, 163–168, 186
   energy budget, 182–183
   laminae, 154–155
   Lowellis Canals, 19, 150
   nitrogen, 182
   oasis craters, 169–170
   oceans, 189
   physical characteristics, 152–153
   polar caps, 159–160
   rainfall, 193–194
   salt, 179–180
   thawing ice, 184–185
   Viking landers, 175, 180–181
   water, 153–154, 157–159, 161, 181
Mass Driver, 126, 129
McKay, Christopher, 182–186
Meadows, Paul, 248–249
Mercury, 21, 233–235
Milankovitch cycles, 79, 79–80, 155–156
Milton, John, 18
mirrors in space, 136–139, 166–167, 236
Moon, 222–233
   Apollo-8 visit, 31–32
   Apollo landings, 223–224, 233
   artificial atmosphere, 225–226, 230–231
   effect on Earth's magnetic field, 62
   effect on Earth's stability, 75–76
   oceans, 228–229
   spinup, 226–228
   tides on Earth, 60
Moravec, Hans, 133
moving planets, 143–146
Mutch, Thomas, 159

NASA
  Ames 1975 study, 28, 116, 163–168, 183, 186–187
  Post-Viking reassessment, 181
  Mars maps, 191
  SOLARES project, 138
Neptune, 21, 68, 241
nitrogen, 56–58
nuclear explosions, 131, 236
nuclear power, 115–116

Oberth, Hermann, 137
oceans, 50–51
O'Keefe, John, 141–142, 226
O'Leary, Brian, 105–106, 232
O'Neill, Gerard, 105, 252–253
oxygen, 55–56
ozone, 42, 47

panspermia, 40
Paradise Lost, 18
photosynthesis, 43–44
planetary engineering, *See terraforming*
Pollack, James, 156, 205
Pournelle, Jerry, 151–152, 207
Project Stormfury, 98

Rawlings, Pat, 212
reverse swingby, 111–113
rings around planets, 140, 141–142
Russian super–irrigation, 96–97

Sagan, Carl, 25–26, 162, 206–207

Saturn
  atmosphere as resource, 213
  moons as resource, 109–110, 169
  ring ice as resource, 110, 111, 150
SETI, 16–17, 26–28, 29–40
shades in space, 139–140
skyhook, 132–133, 175
Snyder, Conway, 179
solar sails, *See light sails*
solar energy, 114
Stapledon, Olaf, 19–21
stellar engineering, 39
Strickland, 184, 185

Terraforming
  environmentalist objections, 256–257
  ethical aspects of, 255–256
  fulcrum concept, 42–43, 90
  historical parallels, 247–251
  Houston colloquium of 1979, 29–30, 94
  human role in, 120–121
  invention of word, 21
  military applications of, 254–255
  motivations for, 244–246, 258–259
  objections to, 251–255
  options, 34–35
  resources, 103–122
  sociology of, 248–249
  tools for, 125
Titan, 241
Triton, 241
tropopause, 46
troposphere, 46–47

Tsiolkovskiy, Konstantin, 18–19, 31, 252

University of Colorado, 16, 182, 188–189
"Mars Project"

Venus, 198–219
    anomalous rotation, 201–202
    as terraforming target, 36
    atmosphere, 198–200, 208–213
    algae seeding, 25–26, 206, 207, 216–217
    climate history, 204–205
    exporting oxygen, 211–212
    importing hydrogen, 212–214
    Last and First Men, 20–21
    oceans, 214–216
    physical properties, 200–201, 202–204
    shading, 141–142, 209–210, 216, 218–219
    sulfuric acid, 208
    terraforming strategy, 216–219
Verne, Jules, 18, 25
volcanoes, 59, 65–66
Vondrak, Richard, 223–226

Warner, Jeffrey, 255–256
water, 51, 68–70
Welch, Stephen, 208, 209
Williamson, Jack, 21–22

Zwicky, 24–25